I0831902

exotic ALIENS

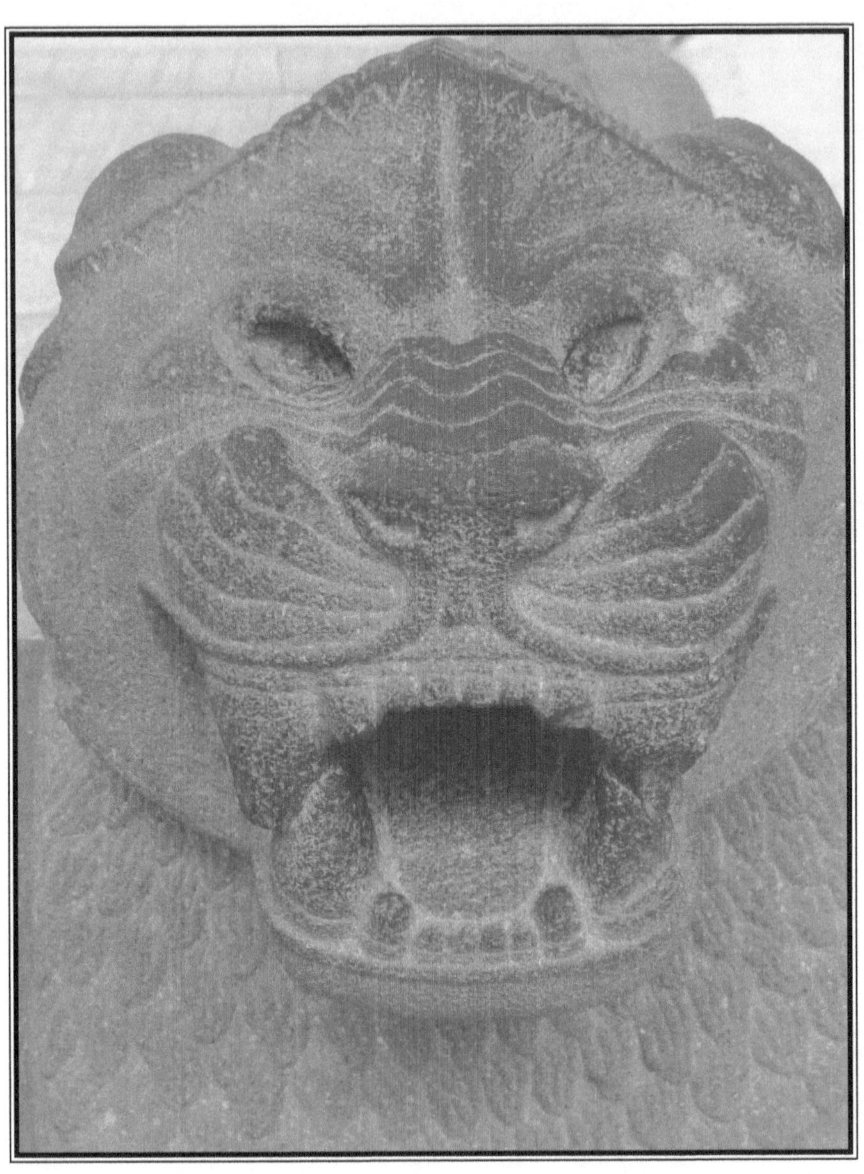

Portal lion from the late Hittite period in the Archaeological Museum, Istanbul (eighth century BCE).

exotic ALIENS

THE LION & THE CHEETAH IN INDIA

VALMIK THAPAR

ROMILA THAPAR · YUSUF ANSARI

Research, Photography and Editorial Assistance
by Hajra Ahmad

ALEPH

For Raj and Romesh Thapar, who inspired challenges.

ALEPH BOOK COMPANY
An independent publishing firm
promoted by ***Rupa Publications India***

This edition published by
Aleph Book Company
7/16 Ansari Road, Daryaganj
New Delhi 110 002

ISBN: 978-93-82277-55-2

1 3 5 7 9 10 8 6 4 2

CONTENTS

Akbar hunting with cheetahs in the sixteenth century CE. Thousands of cheetahs must have coursed the countryside with their masters and keepers. Till as recently as the twentieth century, there were hardly any records of wild cheetahs and most of the cheetahs encountered on record were tame. In fact, for every record of a wild cheetah seen or shot in the nineteenth century, there are at least 2,000 leopards that were recorded to have been seen or hunted. The numbers of cheetahs in the wild are miniscule and never add up.

AUTHOR'S NOTE

This book came about by chance, and had its origin in a remark made by my wife, Sanjna, in my basement library. Although my collection has hundreds of books on India and its fauna that I have acquired over the decades, Sanjna was quite right when she accused me of 'never reading a thing'. Her comment spurred me into action, and I decided I would start reading through all the books I possessed in order to put together an anthology of 500 years of writing on the tiger. That book, called *Tiger Fire*, which seeks to be the definitive anthology of tiger writing, will be published later this year. But in the course of pulling it together, I stumbled upon some facts and an idea took root in my mind.

Why, I wondered, in all the books about Indian fauna I was reading, especially those published in or close to the historical period they were referring to, was there so little about the Indian lion and cheetah? Encounters with lions in the wild were rare, even when they lived in prides, and the same was true about cheetahs which have many cubs and live in a family group for at least 16 months. Both these animals also lived in open grasslands where it would have been easy to spot them, and it was in this habitat that the British did most of their pig-sticking but never encountered lions or cheetahs. Could it be because both these animals were not indigenous to our country or distinctive subspecies, but were exotic imports that had become 'naturalized' over time? The more I thought about the idea and researched it, the more excited I grew about my thesis.

I roped in two other writers—who had the added advantage of being experts on various periods of Indian history—to help me with this project. The first was Romila Thapar, one of India's most distinguished historians, who has written the first chapter of the book, 'The Lion: From Pride to Metaphor', and the second is Yusuf Ansari, who specializes in Mughal history, and who has written the chapters, 'Animals of the Chase' and 'The Shahenshah's Shikar'. Between them and my own contributions to this work, I hope we might have come up with a thought-provoking book on the subject of the lion and the cheetah in this country.

In the process of my research over the last year, I have been hugely inspired and impressed by the book, *The Royal Hunt in Eurasian History* by Thomas T. Allsen. This remarkable work greatly helped me in clearing away the many cobwebs that surround the history of both these species. Similarly, there was Stephen J. O'Brien's work, *Tears of the Cheetah: The Genetic Secrets of Our Animal Ancestors*, that helped me see these animals through the filter of genetics, which substantially enlarged my understanding of them. Divyabhanusinh's authoritative works on the cheetah (*End of the Trail*) and the lion (*The Story of Asia's Lions*) proved to be a treasure trove of information and reference material. Many other titles, especially those on Africa, provided vital facts to strengthen my premise like John Pollard's book, *Africa Zoo Man: The Life-Story of Raymond Hook,* and Nigel Rothfels's *Savages and Beasts: The Birth of the Modern Zoo.*

Besides my co-authors, I would like to thank my wife, Sanjna, especially as she triggered it all, and Hajra for her help with the research, thc visuals and the photographs. My sister, Malvika, and my nephew, Jaisal, helped in more ways than one and my son, Hamir, was a constant source of inspiration. There were many others who provided me with valuable insights into a world gone by. Moreover, I would like to thank all the photographers and proprietors of copyrighted materials for allowing me to use their work. Apologies in advance in case I have made any errors in the credits.

I genuinely hope *Exotic Aliens* creates a bit of a stir in the world of natural history—such debates only help us to better understand the natural world that surrounds us.

Valmik Thapar
New Delhi
1 January 2013

PROLOGUE

Captain Thomas Williamson, the author of the epic book about India's animals, *Oriental Field Sports*, says that in the 1780s, while pig-sticking, one was likely to encounter tigers that had strayed into the open, but never lions, or for that matter, cheetahs. He went on to state that, after spending the last two decades of the eighteenth century roaming the wilderness of India, he believed that there were no lions in Hindustan. For me this was an excellent clue regarding the state of the lion during that time, which was, then, a much debated topic. In his observations of the same period, Thomas Pennant says of lions: 'Those who deny that those animals were natives of India, assert that here was a royal menagerie and that the breed was propagated from the beasts which had escaped.' The debate about the origins and prevalence of lions and cheetahs in India must have been vigorous in the eighteenth century and later but I was intrigued by the fact that it hasn't been talked about much in recent years. In fact, in the twentieth century, most serious observers just took it for granted that lions and cheetahs were indigenous to India.

In his book, *Wild Animals in Central India,* published in 1931, A. A. Dunbar Brander writes that there were no lions in Central India and goes on to say of the cheetah that, 'The Hunting Leopard (*Cynalurus jubatus*) has now almost entirely disappeared from the province without apparent reason, and I only know of three animals having been procured in the last twenty years.' As we will see in the course of this book, it is likely that what Brander referred to must have been three trained cheetahs imported from Africa. Why is it that the comments about lions and cheetahs and their rarity remain unchanged over hundreds of years, especially from people who travelled across India in search of game? Why are pictures of lions as hunting trophies so rare and the few that exist tend to have been taken after 1886 in Gir? Where are the cheetahs in the hunters' bags? After all, according to commentators, hunters were responsible for destroying both these species. As I researched my thesis, I grew convinced that I had stumbled upon the biggest myth perpetrated about these two species in India. Tigers and leopards were everywhere in historical records, whether written, photographed or depicted in other visual forms, but lions and cheetahs were almost invisible. Even on those rare occasions when lions and cheetahs were seen, their behaviour appeared forced and unnatural compared to what I had witnessed of both species in Africa.

As I researched into the prehistoric era (eight to ten thousand years ago) through extraordinary books like Erwin Neumayer's *Lines on Stone: The Prehistoric Rock Art of India*—which reveals the absence of 'maned' felines on rock—as well as through conversations with my aunt, Romila Thapar, and followed the thread of my argument through the pre-Christian, pre-Islamic, Mughal and colonial periods, my conclusion gradually began to coalesce.

In the course of the nearly four decades that I have studied the Indian tiger, I have had occasion to add to the sum of knowledge on that magnificent predator, but this serendipitous book on the Indian lion and cheetah has been stimulating in new and exciting ways. Is there an evolutionary or genetic basis to my thesis that goes beyond the historical record?

▪

The tiger, lion, leopard and jaguar are the four big cats that belong to the genus, *Panthera*. The reason they are grouped together is because of their ability to roar, even though tigers do not roar like lions and leopards. This ability is made possible by the vibration of the thickened vocal folds below the vocal chords in the larynx. This group of cats has probably descended from mammals called Miacids (*Miacidae*), the first true carnivores, which crept around treetops nearly 50 million years ago; at this time they were thought to be the size of martens.

About 20 million years ago, a group of mammals called the

pseudaelurines evolved—these are, in all likelihood, the direct ancestors of all wild cats today. It is believed that two million years ago, the common ancestor to the *Panthera* cats looked like the leopard does today. After a series of genetic comparisons, zoologists believe that the tiger was the first to diverge from the common *Panthera* ancestor, followed by the lion. Because of this fact, lions, leopards and jaguars have more in common with each other than with the tiger. Tigers and lions look very similar when they are stripped of their stripes and this makes it very difficult to prove which one came into existence when. Hence, it has been very difficult for experts to trace the evolutionary path of tigers and lions.

When it comes to separating a species of cat like the tiger or the lion into a subspecies, things that are taken into account are range or physical characteristics. Biologist Andrew Kitchener states: 'Sometimes a species' morphology changes gradually over its geographical distribution so that at the extreme of their range animals may look completely different. This is known as a cline, and even though individuals on the opposite end of the cline may look very different, it is inappropriate to regard them as belonging to different subspecies. There is always gene flow throughout the geographic range, albeit tempered by natural selection for particular morphotypes or perhaps reflecting a pattern of broad hybridization between two populations that differentiated previously in isolation. Sometimes local populations are separated from one another by geographical barriers, such as a river or a mountain range, and they differ in their appearance such that at least 75 per cent of one population can be distinguished from 100 per cent of the other. In such a case, it is possible to recognize two distinct subspecies or geographical races which can be given different scientific names. Finally, in many cases, local populations of morphologically different animals, may meet but have a narrow hybrid zone between the two, where animals of mixed appearance occur. Therefore, there is always some limited gene flow between most subspecies. Although each animal species and subspecies has its own scientific name, science was often not involved in determining their distinctiveness. Many scientific names date back to between the late eighteenth and early twentieth centuries and are often based on a single or a few specimens.' Traditionally, twelve subspecies of lions were recognized based on their geographical distribution, size, shape, and so on (these divisions were based on an analysis of zoo specimens). In the very recent past, eight lion subspecies have been more or less accepted although even within these some are considered invalid. These subspecies are: the Asiatic lion (*Panthera leo persica*), once widespread across Turkey, southwest Asia and India; the Barbary lion (*Panthera leo leo*), originally found from Morocco to Egypt and now extinct; the West African lion (*Panthera leo senegalensis*); the Northeast Congo lion (*Panthera leo azandica*); the East African or Masai lion (*Panthera leo nubica*); the Southwest African lion (*Panthera leo bleyenberghi*); the Southeast African or Transvaal lion (*Panthera leo krugeri*); and the Cape lion (*Panthera leo melanochaita*), now extinct. New research suggests that the modern lion can be divided into three geographical populations: North African and Asian, Southern African and Middle African.

I am sure that in the future this will change again but as far as my historical research is concerned, the Indian or Gir lion is a mixture of some of the above groups.

▪

The cheetah probably evolved in Africa, but some 10,000–12,000 years ago it faced a severe genetic bottleneck in its population resulting in the near extinction of the species; although it survived, a high incidence of inbreeding took place within the remaining population. Till recently, many believed there were at least five subspecies of the cheetah. These are: the Asiatic cheetah (*Acinonyx jubatus venaticus*) that roamed across Iraq, Iran, Israel, Jordan, Oman, Saudi Arabia, Syria, Russia, India and many other countries in the region; the Northwest African cheetah (*Acinonyx jubatus hecki*); the Eastern African cheetah (*Acinonyx jubatus raineyii*); the Southern African cheetah (*Acinonyx jubatus jubatus*) and the Central African cheetah (*Acinonyx jubatus soemmeringii*). The issue of the subspecies of the cheetah is still unresolved as many believe the North African and Asiatic cheetahs are the same but others hold that the Asiatic and African are different. The confusion arises from the fact that genetic variations among cheetahs are very slight and I am sure, with time, this classification will undergo further change. Be that as it may, I believe the cheetah in India has been imported from other countries.

Given the constantly shifting classification of lions and cheetahs based on the evolutionary record, is genetics an effective determinant of whether a subspecies is distinctive or not, especially with regard to the Indian lion and cheetah? Unfortunately, genetic studies on these animals thus far have proved inconclusive and much more work needs to be done to find the answer. What I believe—and this is what my co-authors and I explore throughout the book—is that lions from Persia and Africa were being imported into the country 2,500 years ago (and then on) to meet the demand of Indian royals, and being bred and propagated as court symbols and for hunting; this imported animal was erroneously called the Asiatic lion. The story of the cheetah is much the same but its inability to breed in captivity meant that many more had to be imported. Both species in India were genetic mixtures of animals brought in from elsewhere, and their own inbreeding; their genetic makeup can be best described as a khichdi of genes. This premise of mine requires new research and DNA analysis of both 'subspecies'.

As I am not a geneticist, let me quote Stephen J. O'Brien who, in his eye-opening book, *Tears of the Cheetah: The Genetic Secrets of Our Animal Ancestors* says, '[The] physical traits in Asian lions are manifestations of extremely severe inbreeding in their very recent past. The evidence for our conclusions was encrypted in their genes.' He writes that Asian lions from the zoo or Gir forest had 'virtually zero genetic diversity' compared to the African lion. 'The incidence of malformed sperm was 66% in Gir lions on average, compared to 50% in crater lions and 25% in Serengeti lions. Gir lions had five times fewer motile sperm per ejaculate than Serengeti lions and a tenfold reduction in serum testosterone levels.' The low level of testosterone not only increases the quantity of malformed sperm in a sample of semen, but also affects the development of the mane. 'The Gir lion males had become "feminized" by their history of inbreeding.' Stephen O'Brien was very clear that the DNA of the Gir lions was so similar that it was as if they were all clones or identical twins.

After analyzing samples taken from Asian lions being bred in 38 American zoos, O'Brien stated: 'When we considered the known mutation rate for microsatellites, we calculated that it would take around three thousand years for a severely bottlenecked population...to reconstitute as much new variation as we were seeing in modern Gir lions. The bottleneck that compromised the Gir lions' genetic variation dated back not just one century but three millennia!'

Whatever might have happened around 2,500 to 3,000 years ago to cause severe inbreeding in the lion, this was also the time where I believe the lion as an exotic animal entered India and was propagated by man. O'Brien was also clear that the Asiatic lion being bred in the American zoos was a 'mongrel'; instead of being a 'pure strain', it had two African forebears among the five original founders. As far as I was concerned, this was what connected the Asiatic lion to Africa.

O'Brien believed that the Gir peninsula (which is the only part of the country where the Asiatic lion is found) was like an island 2,500 years ago and the isolated population of lions in it suffered from severe inbreeding. Many others have agreed that this area along with Kutch was once an island, as were parts of Kathiawar. Romila Thapar states: 'Even if this was lion country, there seem not to have been lions here. If there were lions here, why do they not occur on Indus seals, since the area had many Indus cities and settlements, such as the city of Dholavira on the edge of the Great Rann of Kutch? It would seem that there were no lions in this area during the period of the Harappan cities, around the mid-second millennium BCE.' If there were no lions 4,500 years ago and they remained rare in the periods that followed, be it the Islamic, Mughal or British periods, then it must have been an exotic import all along as there is no gradual decline in the species.

I believe that the lion came to India just before or with Alexander's invasion of India around the third or fourth century

BCE. It was from this time onwards that lions were bred, either in designated hunting parks or royal zoos, just like the Gir lion was bred in the late nineteenth century. The breeding of lions in ancient times was for the hunt, or to keep in menageries or in the court of the king as symbols of royal power. A recent paper by Barnett, Yamaguchi and Cooper entitled 'The Origin, Current Diversity and Future Conservation of the Modern Lion (*Panthera leo*)' states: 'Over its natural range the modern lion may be evolutionarily subdivided into three geographic populations: North African-Asian, southern African and middle African, determined by expansions and contractions of mid latitude deserts and low latitude forests. The former two are characterized by relatively simple mtDNA haplotype structure, and the latter by a diverse haplotype mosaic. The North African-Asian population is separated from the Sub-Saharan African populations by the vast Sahara Desert, and its skull is morphologically distinguishable from those of Sub-Saharan lions (Hemmer 1974), suggesting its distinct taxonomic status.'

The interesting aspect of this piece of research is that these scientists put the North African and Asian lion into a single subspecies. In other words, there was just no 'separate' Asiatic lion. László Bartosiewicz reiterates this fact by saying: 'Modern lions are closely related. Craniological traits of the now extinct North African Barbary lion (*Panthera leo leo linnaeus*, 1758) and Asiatic lions (*Panthera leo persica meyer*, 1826) are very similar. There must have been a contiguous population inhabiting North Africa and Asia.'

Even though the genetics of the species is complicated and necessarily the preserve of scientists, my research into some of the key findings in this area only strengthened my suspicions that the lion in India was an exotic alien.

Where the cheetah in India is concerned, we should track back to O'Brien and his basic premise, where he argues that the biggest bottleneck in the history of the species was 10,000-12,000 years ago when there were major changes and upheavals across the planet. He writes that the cataclysm that claimed saber-tooths, mastodons and so on almost decimated the cheetah as well; it was that 'narrow escape from extinction' that left its genetic mark of uniformity on the survivors. If, as O'Brien postulates, the major population of the cheetah in Africa itself is narrowly genetically diverse because of a 'population bottleneck' the species passed through millennia ago, I think it is extremely unlikely that a genetically distinctive subspecies could have existed in the subcontinent. We will explore this at greater length in my chapter on cheetahs.

I believe there was never an 'Asiatic cheetah'. Rather, this animal was an imported royal pet that escaped into the wild on several occasions and may, at times, have created small feral populations. The evidence that my co-authors and I have unearthed in the course of our research and study has duly underlined the basic premise that I began this book with—that the Indian (or Asiatic) lion and the Indian (or Asiatic) cheetah are not distinctive subspecies but are exotic aliens that live (or lived) in the land of the tiger. For me, this fact was made amply clear as I ploughed through hundreds of books and research papers on both the animals. The numbers just didn't add up—I found fewer and fewer encounters with lions and cheetahs the more I read. Descriptions of both species were rare, and both seemed to lack a 'tradition' in India. Tigers and leopards were all over the place; lions and cheetahs in comparison were barely visible in the pages of history. Till the end of the nineteenth century, there were more lions seen in royal menageries than in the wild and hundreds (if not thousands) of cheetahs on leashes were recorded as being readied for coursing on bullock carts, but not even a handful were encountered in the wild. This book is the story of our search and raises serious questions about the ancestry and place of the lion and the cheetah in their natural state in India.

If for some reason the premise I have put forward is proved to be wrong, I will be happy to be corrected. The only thing that matters is that the provenance of these two amazing animals is conclusively proved. I hope that the book provokes even more research into these incredible animals and their history in India.

THE LION: FROM PRIDE TO METAPHOR

by ROMILA THAPAR

The lion pillar at Vaishali in Bihar (third century BCE).

Next page:
The lion on the pillar at Vaishali. This is the beginning of the first 'lion art' in India in the third century BCE, about 100 years after Alexander's campaign in India.

I am not a specialist in the history of wildlife in India, let alone the lion. My interest was initially marginal, whetted by visits to sanctuaries in India, East Africa and the Amazon, with some reading on these areas, but more so by conversations with my nephew, Valmik, essentially on the tiger. The lion first entered my horizon when I studied the Mauryan sculptures of the animal where it forms the capital of pillars. I felt that these reflected less familiarity with the animal than we have assumed. The contrast is particularly striking when compared to Assyrian and Achaemenid representations. I expressed my doubts in conversations with Divyabhanusinh when he was writing his fine book, *The Story of Asia's Lions*. Divya was convinced that the lion was indigenous and an animal of the Indian wild. We agreed to disagree! In my recent conversations with Valmik, I have been persuaded to pen my doubts. Therefore, I am attempting here to explore some aspects of the question as to whether the lion was indigenous to India or whether it was brought to the country at some point in history, after which it proliferated up to a point, but more importantly, it became an idiom of majesty, prowess and power.

THE QUESTION

The lion is among the more extraordinary animals of the natural world given the infinite ways in which humans have related to it. At the most obvious level, it is the principal predator of the grasslands, the savannahs, where many other four-footed animals are its prey. It can also be a man-eater when its supply of food diminishes. But at multiple other levels—psychological, magisterial and in fantasy—the evocation of the lion by human societies has taken myriad forms. The extraordinary part of this evocation is that, in some places, it is a reflection of the presence of lions in proximity to human settlements, and in yet other areas, there is only the appropriation of the symbolic meaning of an association with the lion, where the presence of the lion may have been largely known only from its ceremonial function or even from hearsay as an exotic animal. There is, therefore, in most Eurasian cultures, the perception of either its physicality as part of the natural landscape, and/or its being an icon of power and belief.

This raises the question of determining the original habitat of the lion, a question that can, in fact, only be answered by scientists who specialize in the study of the species as part of the natural world. Non-scientists can use historical and literary data to assess the forms in which the lion is evoked and hopefully this can become less opaque with further investigation. Keeping this in mind, we may ask whether the Indian subcontinent, either in its entirety or in limited areas, was a natural habitat for the lion. This becomes an obvious question given the contrast with the tiger whose indigenous habitat being India, apart from elsewhere in the East, cannot be questioned.

I am concerned here with references to the lion and its representation in India from early times, and in assessing whether it was regarded as an indigenous animal or an exotic that gradually either became naturalized or continued to be imported. A similar question has been asked of the cheetah, with perhaps less conclusive evidence of its presence. In attempting to find answers, it might be salutary to look at the presence of the lion in places where it was known to be indigenous and familiar. This involves some comparison with contemporary societies of early times. This brief investigation is not intended as a history of the lion in early India as that would require a different analysis.

The intention is not to prove or disprove India being the natural habitat of the lion or the cheetah, but rather to consider when their presence came to be recorded and how they were perceived.

Inevitably, this leads to many questions. Are lions always present in the same habitat or does their habitat change? Are there reasons for their not surviving in areas where they were earlier known to live? Why is the lion hunted in some societies and perhaps not so in others, or at least not so frequently? What does the representation of the lion symbolize? Does this symbolism change? How do the symbols relate to the culture where they are being used? Does this relationship reflect an animal that is merely familiar, or is its presence essential to that culture?

Prehistory

Even partial answers to these questions takes one back to a remote past. Fossil remains of lions from East Africa date to over a million years ago. Their earliest representations, sometimes from about 30,000 years ago, come from prehistoric cave paintings in various parts of the world. The Chauvet-Pont-d'Arc Cave in France, discovered in 1994, revealed many animal murals. Lions, horses, mammoths and rhinoceroses are painted on the walls of the earlier caves and these gradually change to smaller animals in the later caves. The most dramatic is the panel of lions. In historical times, the frequency of the lion occurs in more southerly areas. Prehistoric cave sites in Europe have drawings of animals that are now absent in that area. That these animals may have been hunted to extinction may not be a sufficient answer, since not all the animals depicted were those that were hunted. Did the coming of the Ice Age so change the environment that the lion could no longer survive in Europe?

As to why they were depicted in these prehistoric murals, one suggestion by James Frazer, and endorsed by Joseph Campbell, was that the drawings could be a form of 'sympathetic magic' where the representation of the object is believed to ensure a person's access to it or, better still, the capturing of its essence. Hunter-gatherer societies that are associated with prehistoric murals had shamans, who were their specialists in ritual. A shaman would sometimes go into a trance in the depth of a cave, and in that state, paint murals often depicting the most ardent wish of the community. This wish would naturally be for success in the hunt and gaining control over the object of the hunt. So, if large animals were the object, they were painted. The paintings were linked to the magic of the hunt, the intention being to replicate the success of the expedition as painted in the rock shelters and caves. Success, of course, also involved a propitiation of the deity that controlled the hunt.

However, rock art in India of prehistoric times, found at many sites especially in Central India, does not provide paintings of the lion. For instance, Bhimbetka rock shelters in Madhya Pradesh, the site which is virtually a museum featuring many animals known to India over centuries, does not seem to know the lion.

The Icon of Power and Protection

As the protagonist amongst the animals hunted in many areas, the initial association of the lion seems to have been with prowess. Grappling with a lion was a test of heroism, where heroism was the most respected, crucial quality amongst members of most early societies. Animal herders were well aware of the threat from lions. The Masai in East Africa, living in an area known to be the habitat of the lion, have a rite of passage for a male adolescent entering manhood where he has to overcome a lion in single combat, carrying only a spear. A group of eligible young men undergo this test and the one who first strikes down a lion is chosen to be honoured. Acknowledged as a great hero, he becomes eligible for being elected as the great chief.

The germ of this idea is seeded in many cultures of early times. The sphinx in Egypt suggests the pharaoh/man imbibing the qualities of the lion. A more direct and frequent motif, which is repeated in many parts of the Mediterranean world and in West Asia, is the depiction of the hero combating a lion. Was this just the common experience of all heroes in places where there were lions or does the idea go back to the ancient Mesopotamian myth of Gilgamesh? What is striking is that, in ancient Egypt, the representation is not of violence but of power. The prowess and majesty of the lion is sometimes combined with the non-terrestrial, aerial creature that carries the same symbolism—the falcon, which is also worshipped in Egypt. With the introduction of winged lions, there is a suggestion that perhaps the two were conflated.

Prowess is not the only association with the leonine and the male hero is not the only one to be honoured. In Egypt, the lion-headed goddess, Sekhmet, was the goddess of protection and creation and was associated with the sun as a deity. This linked symbol occurs many centuries later in Central Asia, where it was used by Mongol and Turkish conquerors. The counterparts of Sekhmet from elsewhere, such as the goddesses Ishtar, Inanna and Anahita from West Asia, are shown either standing beside a lion or seated on one. Curiously, despite the close links to the sacred, the lion was also hunted for sport.

LION HUNTING

That lion hunting was required of kingship is evident from many Egyptian and Mesopotamian sources. Interestingly, two types of hunting are described. One was the hunting in the wild of lions, elephants and wild bulls. The wasteland north of Memphis in the upper Nile delta was associated with such hunts. The other was hunting within specially created 'hunting parks'. These were large acreages of selected areas, not too distant from the capital, that were cordoned off and where kings could hunt at their convenience. Their proximity to the capital facilitated the frequency of hunting expeditions, even if only to the parks. The hunters, mainly kings, could choose to use beaters who drove the animals into a smaller area, thus facilitating the hunt.

The numbers of lions killed, as listed in Egyptian and Assyrian sources, are enormous and may well have been exaggerated to underline the king's prowess. The location of such parks could have been where big game was not easily available and had to be either coaxed into the park, or literally transported there in cages on river barges, and then released to roam in a limited area. This is shown in some Assyrian low relief stone panels (such as those displayed at the British Museum). The idea of the hunting park which made hunting easier became fashionable among kings, perhaps because it was instinctive for royalty to want to declare its might. Soon, such parks became characteristic of a well-established kingdom. And as with all such activities, they were the cause of immense competition.

In Mesopotamia, the best hunting areas were in the land between the two rivers—the Tigris and the Euphrates—where the major cities were located, as well as the hunting parks. Projected as gardens of paradise, their mythology may even have influenced the biblical idea of the Garden of Eden—of course, without the hunting of lions! But representations of kings hunting lions are depicted on cylinder seals, stone relief carvings and in murals on temple walls, where lions are guardian figures. The Assyrian kings both massacred lions and appropriated them as icons of power and protection, in a seemingly contradictory way. The great codifier of laws, King Hammurabi, made lion hunting the sport of kings. Lions were trapped and brought to the hunting parks near the capital. Once there, the animals were protected by strict game laws, as only the king had the right to hunt them. Tiglath-Pileser I claims to have killed 920 lions, Ashurnasirpal II killed 370 and Senacherib had caged lions released in the hunting parks for the royal hunt. The different stages of the hunt are depicted in the low relief panels on palace walls, now regarded as masterpieces of Assyrian art. The lions could be caught in the wild or else bred and tamed in captivity before being let loose in the hunting parks.

The image of the hero in combat with the lion goes back to an early myth of Sumer—that of the hero Gilgamesh of the third millennium BCE. The graphic description of his dream in

Alabaster relief of a lion hunt from Assyria dating to the ninth century BCE. There is a remarkable difference in the depiction of lions from Assyrian sources as compared to others.

which he combats lions on all sides becomes a familiar motif. The murals of a later millennium that depicted the king in combat with and hunting lions had, by now, both a political and religious purpose. The procedures in the hunting parks were virtually a staged ritual. Animals emerged from cages in the park, and there is also the representation of a handler with a mastiff as part of the lion hunt. The Assyrian kings were making a political statement in appropriating the symbolism of the lion from the preceding dynasties of Sumer and Akkad in Mesopotamia and from the lion-hunting practices of Egypt. The muscular power of the lion and the realism with which it is sculpted makes the intention quite clear. The Assyrians knew their lions. The panels show the lions, open-mouthed and roaring, generally on their hind legs, their neatly arranged manes extending below the forelegs, and the body exuding both the feline strength of the big cat and the tenacity of the predator. Many sites, such as Susa, Khorshabad, Balawat, Lachish, had such panels in their palaces, but the finest examples come from those at Nimrud and Nineveh.

Assyrian relief from Ninevah of a wounded lion dating to the seventh century BCE.

The Spread of the Cult of the Lion

Westwards

The cult of the lion, if one may call it that, with what seems to have been its epicentre in Egypt and Mesopotamia, subsequently spread both westwards and eastwards from the second millennium BCE. In Greece, the famous gate of the royal enclosure at the site of Mycenae, of the late second millennium BCE, was shown as guarded by lions, suggesting a West Asian idea. Perhaps the most enduring idiom was that of the Greek hero, Hercules/Heracles. He was required to perform twelve feats, each an almost superhuman act, one of which was to kill a lion—perhaps in imitation of Gilgamesh. This he proceeded to do by strangling it.

The lion was now part of Greek legend although it may not have populated the landscape of Greece to any noticeable degree. The family of Alexander claimed descent from the lion which did not deter Alexander from seeing himself depicted as hunting lions.

Further west, Etruscan art in Italy borrowed the lion motif from the Orient and used it in a number of small ornamental objects, such as broaches, belt clasps and jug handles. The lions are depicted in a stylized manner, doubtless imitating eastern originals made familiar through trade with Egypt and the eastern Mediterranean. In the neighbourhood of the eastern Mediterranean, Jewish mythology focused on the lion as the symbol of the Messiah to come, zeroing in on the concept of the Lion of Judah—a concept that played a prominent role in later Ethiopian culture. With the rise of Rome and the boundaries of the Roman Empire extending to North Africa and further east than the Mediterranean, these contacts were intensified and activities became more intercultural. Egypt in the third century BCE had connections with Rome and, to a lesser extent, with West Asia and India. Ptolemy II Philadelphus of Egypt is mentioned in an inscription of the Mauryan king, Ashoka. Elsewhere, he is said to have hunted with golden spears and used Indian hounds. These possibly were gifts from the Mauryan king although they are not included in the lists of the gifts exchanged.

Ashoka, in one of his inscriptions, mentions five contemporary kings of West Asia and the eastern Mediterranean whom he refers to as *yonarajas,* kings of the Yavanas/Greeks. The five were closely related so it is not clear whether Ashoka sent missions to each separately. Three among them sent ambassadors to the Mauryan court at Pataliputra (modern Patna), but there is no certainty that they were at the court. The neighbouring kingdom to the west was ruled by Antiochus II Theos, the grandson of Seleucus I Nicator, one of the generals of Alexander ruling in West Asia. Seleucus had a treaty with Chandragupta, so there were friendly connections with the Mauryas. He sent Megasthenes to India who left an account of the country, which unfortunately has not survived except as lengthy 'quotations' in various later Greek texts, where the 'quotations' do not always agree. The authenticity of some of the details in the account has been a matter of debate, as is often the case with such descriptions. The envoy from the Ptolemy king was Dionysus and the third envoy was Deimachus, a Greek from the Seleucid Empire. Mention is made of the Indian king requesting sweet wine, figs and a Sophist, of which only the first two could be sent. However, no mention is made of a request for large animals. Hunting dogs from India are said to be particularly fierce, so much so that Aristotle attributed this to the interbreeding of dogs with tigers. Their strength was demonstrated by pitting them against a lion. Hunting dogs that were the like of leopards are described in Chinese sources as coming from Samarkand in Central Asia in the form of tribute. They were more likely a gift, but Chinese record keepers at the court had a proclivity for treating all gifts as tributes! Hunting dogs may have been a novelty in China, but it is puzzling as to why they would be so highly prized in Egypt when the cheetah was found close by to the south.

If they came from India, the hunting dogs could either have come by ship or over land through West Asia. Some sources refer to Syria as an entrepôt for goods coming from the East. Maritime contacts via the Red Sea became regular a couple of centuries later. Mention is made of peacocks, pythons and tigers sent as gifts from Indian rulers, but not lions. The Deccan (also Dachinabades, Dakshinapatha) was said to be the area where leopards, tigers, elephants, huge serpents, hyenas and varieties of monkeys were found in abundance, but the lion is not listed. By the second century BCE, Roman landowners were cordoning off areas to be used as hunting parks which had then become the fashion. Caesar's gladiatorial animals, especially lions, came from Libya, Syria and Tunisia, the latter attested to by excavations near Carthage. These were provinces of the Roman Empire. The entertainment and spectacles provided for the Roman public included exotic animals among which the lion was prominent, the tiger far less so. (As a contrast, the later animal collections of the emperor Kublai Khan included many tigers, more easily transported from India to Central Asia. The Siberian tiger may have been difficult to acquire and other tigers in Central Asia were alien when compared to lions.) Many other animals from various parts of the (then) known world came to Rome either as exotic gifts or for performances. Bread and circuses was one way of keeping a restless populace quiet, so the more exotic the animals, the greater was the diversion from the problems of daily life. Somewhat later, the Roman emperor Hadrian discontinued the gladiatorial combats and instead organized hunts, presumably projecting himself as what he thought was an Asian style monarch. In Europe generally, the lion becomes prominent in heraldic design, after the Crusades of the early second millennium CE.

Top: *Relief from Sanchi (first century BCE). Winged lions caught the fancy of many cultures.*
Bottom: *Achaemenid stone relief of the fifth century BCE, as a rendering of the idea of the sphinx. It is a guardian deity and was linked to Ahura-Mazda, the Zoroastrian deity.*

Eastwards

Eastwards in Iran, the impact of the lion was strong and visible. The Achaemenid ruler of the sixth century BCE, Cyrus, understood what his Assyrian predecessors had maintained, that lion hunts could also be used as a method of training soldiers and officers for war. But lion hunting for this purpose had to be a controlled hunt and not a free-for-all in lion habitats. Therefore, some lions would be specially brought for this purpose and a strategy for the hunt worked out. This can perhaps be seen as another form of 'sympathetic magic'—capturing that which is being hunted, where the lion hunt becomes a surrogate for battle.

A collection of remarkable gold objects was found by the river Oxus, and has been named the 'Oxus Treasure'. The objects came from Achaemenid settlements in the area and from trade. Among them were vessels, gold plaques, the model of a chariot-and-four, rings, coins, figurines, winged lions and cylinder seals. But the prize object was a gold scabbard that portrays a royal hunt, probably of the time of Darius, a close successor to Cyrus. Horsemen are shown hunting lions and the affinity with the Assyrian low relief panels in the depiction of the lion is evident. Many objects of gold used in the palace, ranging from bowls to bracelets, were decorated with a lion motif.

Darius brought the lion to the fore as the animal of majesty, continuing the tradition of the Assyrian kings. He decorated the palace with scenes of lion hunts and had seals cut depicting the same. Winged lions were also introduced. This may have been in imitation of earlier Egyptian models or there might have been an attempt to associate the lion with the symbol of Ahura Mazda, the great god worshipped by the Zoroastrians, whose iconography had wings. The Achaemenids were Zoroastrians. The personal seal of Darius showed the king hunting a lion from his chariot, with the statement: 'I am Darius the king...'

In the constellation of stars, importance was given by Greek and Iranian astronomers to the one named after the lion, Leo, the fifth sign of the zodiac. The constellation was seen as outlining the shape of a lion and in Greek astronomy was linked to the Nemean lion killed by Hercules. It appears at the Spring Equinox, and its position was used to calculate the Babylonian Calendar as well as the agricultural calendar. Constellations and asterisms carry names taken from mythology. The Greek names have Mesopotamian parallels which would have preceded the Greek and the Greeks took the names and information from earlier astronomy. The contact was most likely made through the mixed populations in western Turkey. Leo was crucial to the belief and spread of astrology in the Mediterranean world and in West Asia. The name 'Leo' was to become a common name to a series of Roman Catholic popes as also some rulers of European kingdoms.

The Achaemenids had inherited the Assyrian idiom of the lion as symbolic of majesty, power, authority and as a source of protection through astrological links. They, therefore, deemed it crucial to the events of the future. The lion had a much stronger presence in Achaemenid Iran than it had in Greco-Roman culture. Unlike the Assyrian and Egyptian kings, the Achaemenid rulers did not boast of the numbers of lions they had killed. Perhaps there were fewer lions in Iran and the ones used in the hunting parks for training the army had to be imported. Nevertheless, the lion as a symbol is prominently visible in the graphics at the capital city of Persepolis. When the foremost epic of Iran—Firdausi's *Shahnameh*, composed in the early eleventh century CE—looks back on the great heroes of antiquity, Rustam and Bahram Gur, it describes their great feats including combating lions. It is this visible presence of the symbol that appears to have been assimilated but reconstructed in the Indian concept and representation of the lion.

Top: *Stylized Sumerian seal with lions, possibly from the third millennium* BCE. *Probably a depiction of the Gilgamesh myth.* Bottom: *A seal from the Indus Valley civilization showing a hero in combat with tigers, third millennium* BCE. *This might have been a version adapted to the Indus region where tigers were the more familiar animal.*

The Lion in India

In comparison with Egypt and West Asia, the lion does not take centre stage in India; in fact, it seems to have been initially absent and was a late arrival, judging by the existing evidence. It is not among the representations of animals, whether in realistic forms or as icons, for many early centuries. Among the animals depicted on seals, on pottery and in three-dimensional forms in the Harappan cities, the lion is conspicuously absent, whereas the tiger, the bull, the rhinoceros and even the mythical unicorn are conspicuously present. This is puzzling as the lion was already a symbol of significance in Mesopotamia with which the Harappans had trading contacts. Furthermore, some of the areas where the Harappans were settled (such as Gujarat) were areas where the import of lions has thrived in recent times, even if in hunting parks, since the ecology suited lions. The Harappan civilization covered a vast area of northern and western India, extending eastwards as far as Punjab and the Upper Doab and south to the borders of Maharashtra. This was a big enough area to have hosted lions in the wild.

The absence is underlined by a specific object: a seal with a scene familiar from West Asia depicting a hero grappling with a lion on either side of him. This scene is repeated on a Harappan seal, the difference being that both the animals are tigers and not lions. It was almost as if the story was adapted to local conditions and the lions, being locally unknown, were replaced with the known and familiar animal—the tiger. In fact, the tiger is shown in various contexts, some of which have been interpreted as depicting a myth. The iconic meaning of the lion had clearly not been appropriated by the Harappans.

Many centuries later, the first of the Vedic texts, the *Rig Veda*, records the reverse condition. It refers to the thunderous roar of the lion, *simha*, which lives in the hills and is a beast that kills, but the *Rig Veda* does not know the tiger, *vyaghra*. A possible explanation could be that the geographical context of the earliest hymns of the *Rig Veda* was beyond the far northwest of India and impinged on northeastern Iran and the Oxus valley. The *Rig Veda* refers to the ensnaring of a lion (10.28.10) which was probably a joint activity of a clan, although in later centuries it is associated with hunting parks. If there were hunting parks in this region they would have been pre-Achaemenid, since the *Rig Veda* dates to the late second millennium BCE, whereas the Achaemenids ruled in the first millennium BCE. There is no evidence of such parks in India in the pre-Mauryan period. The Zoroastrian sacred text, *Avesta,* was composed approximately in this area and there was a close linguistic and cultural relationship between the *Avesta* (in Old Iranian) and the *Rig Veda* (in Old Indo-Aryan). Some clans mentioned in the *Rig Veda* may have been migrants from Iran. Was the lion remembered by them from earlier times even if it was not met with in the Punjab where part of the *Rig Veda* was composed?

If the Oxus Valley (the region that was to be called Balkh or Bactria) lying to the north of the Hindu Kush and the Pamirs, was a habitat of the lion, then this might explain why Harappan settlements in the Pamirs would not have known the lion—the high mountain ranges may have been a barrier. An additional barrier was the Indus River, torrential in its upper reaches but equally formidable lower down. Unlike tigers, lions are known to loathe being in water and, therefore, do not swim across large rivers. Taking them across would require bringing caged lions in boats. The tiger is mentioned in the Vedic texts composed subsequent to the *Rig Veda* in the first millennium BCE, and especially in the *Atharva Veda*, composed in the eastern Gangetic plain where the tiger was a familiar animal of the forest. But the lion also finds mention in these texts. Was this a reference from ancestral memory or does it reflect a closer knowledge of the lion? If the *Atharva Veda* was composed in eastern India, as has been argued, then this was not a likely habitat for the lion.

The tiger is the more frequently mentioned animal in early Indian literature. In the *Mahabharata,* prowess is associated with the tiger and strength with the bull. The hero is frequently referred to as a 'tiger among men' or is compared with the bull rather than the lion; occasionally, the hero is said to have the gait of the lion. The royal seat, however, is the *simhasana* (the 'lion seat'), as was the case in West Asia as well. Yet, when the forest on the Yamuna, the Khandava Vana, is set on fire in order to clear it in preparation for constructing the city of Indraprastha, both tigers and lions are said to have died in the conflagration. These would appear to be animals living in the wild as there is no reference to its being a reserved forest or hunting park. It is plausible that references to tigers and lions together had by now become a literary convention while describing a forest. There is, however, no suggestion that tigers and lions may have occupied different ecological habitats, as they are wont to do—lions in grasslands and tigers in tropical forests. It is almost as if all generalized descriptions referred indiscriminately to lions and tigers as part of the scene in the wilderness and the forest, as indeed they referred to elephants, wild bulls, leopards and many other animals as part of the forest community.

Nevertheless, there is a remarkable statement in the *Mahabharata* that reads:

> The forest has tigers and it should never be cut nor should the tigers be chased away from the forest. Not living in the forest is death to the tiger and in the absence of the tiger, the forest is annihilated. The tiger protects the forest and the forest nurtures the tiger. (5.29.47-48)

This primary interdependence does not refer to the lion.

The lack of familiarity with the tiger in West Asia is a contrast to the presence of the lion. The most characteristic animal from India that intrigued the Mediterranean world—which thought it to be strange—is described in a sixth century Greek text written by Ktesias, a Greek physician at the Achaemenid court in Iran. The text—called *Indika*, as all such texts were called—purports to be a description of the landscape, the people and the animals of India. Ktesias never visited India but based his description on accounts related to him by Indian envoys and merchants visiting Iran, who seemed to have done so frequently enough to create an interest in all things Indian. These merchants were part of a triangular trade between Bactria, Iran and India. Like many such early texts, there is a hint of reality wrapped in layers of fantasy. The animal he describes at length and with much wonder, is the one he calls the *martichora,* perhaps meaning 'the man-eater'. The word is thought to be derived from the Old Iranian words, *martiya* (man) and *khvar* (eat).

> It has three rows of teeth, human ears, and light blue eyes like a man's. It has a tail like a land scorpion on which there is a stinger more than a cubit long… If approached it stabs with a stinger...

In Greek, the *martichora* was said to have meant man-eater, and Ktesias claims to have knowledge of many of these 'creatures' living in India where 'natives kill[ed] them by firing arrows while riding elephants.'

It is thought that this was a garbled reference to the tiger which was the animal most commonly associated with India, and which apparently had been gifted to the Achaemenid king by an Indian ruler. Had it been a lion, Ktesias and the Persians would have recognized it as such. Ktesias lists many of the important animals of India, some of which he describes. The lion finds no mention among these.

The Representation of the Lion in India

Relief and capital from Sanchi that represent a later and even later attempt at replicating the Ashokan pillar but where the lions are less leonine.

The lion leaps into prominence with Ashoka Maurya in the third century BCE, and interestingly, less as the king of the beasts in the wild than as the symbol of supremacy and authority. The earliest representation of lions is in the capitals of the pillars of Ashoka. There are a couple of sejant, seated lions, as well as the now familiar form of four lions addorsed, back-to-back, each facing a cardinal direction. These sculptures have raised many questions related to whether the lion was a familiar animal on the Indian scene or better known as an icon. Do they reflect observations of lions in the wild or tame lions, or are they stylized lions whose iconographic meaning has been adopted by Mauryan royalty from the symbolism of the lion in Achaemenid Iran with its ancestry in the Assyrian?

The four lions from the Sarnath pillar are the ones generally discussed. The form is sometimes linked to clusters from West Asian locations, hinting at a possible Iranian prototype. What weakens the link is also the fact that the connotation is entirely different: the four lions were surmounted by a wheel—*dharmachakra*, the wheel of law. This particular rendering of the wheel is an icon of the Buddhist *cakkavatti*, the universal monarch who rules, not by conquest, but by observing the laws of social ethics. The totality of the imagery immediately brings to mind the proclamation of the Buddha's teaching, which when recorded drew on the simile of its being heard as widely as was the roar of the lion—the *siha-nad*. Further, his initial teachings at Sarnath, the 'Cakkavatti Sutta', which came to be known as the Buddha's *dhamma/dharma,* set in motion the wheel of his new philosophy that was to challenge Vedic Brahmanism, and was, therefore, viewed, and further justified its being viewed as the roar of the lion. This was different from the Persepolitan projection of the lion as royal and majestic. The Ashokan lion was an encapsulation of the power of Ashoka's interpretation of *dhamma*, as expounded in the edicts also inscribed on the pillar, and was being used less as a representation of the animal and more as a metaphor for the teaching. As an animal in its natural habitat dependent on violence for its survival, it was not exactly the obvious choice for the teaching of *ahimsa* (non-violence). A similar meaning was probably meant to be conveyed by the seated lions on the gateways at Sanchi that were built later, where they were the icons guarding the *dhamma*. This becomes obvious if we remember that there was originally a wheel representing the *dhamma* which surmounted the four lions, as on the later pillar from Sanchi, although in the latter the lions look even less leonine than those at Sarnath.

The four lions at Sarnath are not depicted as fierce and powerful animals as in the Assyrian and Achaemenid forms, but as relatively gentle. Therefore, the question remains whether the sculptors saw them in the wild, as part of a royal hunt, or whether they saw them as the animals of a royal menagerie. Their heavy manes end at their forelegs, and are manicured in neat arrangements reminiscent of the Assyrian renderings, rather than the open hair of a natural mane. They lack the feline body or the muscular strength of the West Asian lion. It would seem that the rendering of the lions is, therefore, not of great consequence, and the stylization conforms more to an idea than to the natural form of the lion in its indigenous habitat.

The Achaemenid and Mauryan Contact

A century ago, it was argued that the Ashokan capitals were based on Greek prototypes and heavily influenced by Hellenistic forms. Greek prototypes of lion sculptures would not have been available to Mauryan sculptors, and the argument that Greek sculptors were brought to India to sculpt the pillar capitals has little to recommend it. However, the Iranian models of pillars were closer to India. As to the sejant lion, seated on its hind legs, there is only one way in which it can be depicted, and the embellishments are dissimilar in the Greek and Indian forms. There are, however, some Persian touches in the Sarnath capital although it is distinctively Mauryan. The sculptured capital as an entity capping the pillar may be suggestive of an appreciation of Persepolitan forms, but again, it was Mauryan in concept and execution, barring the occasional design of palmettes and acanthus leaves. The crafting of the elephant, bull and horse on the abacus, the band below the capital, is a contrast to what is above it. The realistic representation of the animals on the band and their closeness to earlier Indian forms is striking. The lion alone has a hint of unfamiliarity.

The characteristic feature of civilizations is that they do not evolve in confined, bounded, sealed territories, as was once thought. Civilizations are by their nature, porous and where they are contiguous, their porosity increases. One does not have to look beyond proximities to see similarities. This was the case with the Achaemenids and the Mauryas. In the period prior to the Mauryas, the Achaemenid Empire extended as far as the Indus River and the *satrapies* (provinces) included Gandhara (in the Indo-Afghan borderlands) and Sindhu (the Indus region). The successor to the Achaemenids in the area, the Seleucids, ceded the territories of eastern Afghanistan, as far south as the Makran coast, to the Mauryan emperor. This accounts for the ease with which ideas and artefacts crossed boundaries. Among these were: the style of royal inscriptions and their being engraved on rock faces and stone pillars; the use of the Kharoshthi script for writing Prakrit which drew from the Iranian Aramaic script, although its use was limited to northwestern India; envoys and merchants who travelled between northern India and West Asia, Egypt and the eastern Mediterranean; trade between these areas promoting the prosperity of their cities. Not surprisingly, some Achaemenid forms and ideas were assimilated by Indians. As in all civilizational dialogues, these were reformulated and adapted to new contexts. If we can say that the Mauryan lion and what it connotes was among these, then the use of the lion as a metaphor suggests a fascinating lineage from the Egyptians in Africa, to the Assyrians in Mesopotamia, to the Achaemenids in Iran, to the Mauryans in India. Did the lion also travel from Africa to West Asia to India?

▪

However, the idea of the lion in India was not fantasy. The question is whether it was, as was the tiger, an animal indigenous to the Indian wild, or was much of the familiarity with the lion based on seeing it in near captivity. Sanskrit and Prakrit texts list lions along with many other animals as inhabitants of the forests, but what is not always clear is whether they are in the wild or in reserved forests.

There are accounts of India in Greek and Latin by non-Indian commentators that refer to lions, sometimes in somewhat puzzling ways. Most of these descriptions claiming to refer to Alexander's campaign in India were written a few centuries later. These are not eyewitness accounts but descriptions based on earlier accounts, current hearsay and possibly some contemporary situations. They cannot, therefore, be taken literally in every case. Thus we are told that specially bred dogs are said to have been used for hunting animals, including the lion. This description more likely refers to hunting parks.

> I reckon Indian dogs amongst wild beasts for they are of surpassing strength and ferocity and are the largest of all dogs. This dog despises other animals but fights with the lion, withstands his attacks, returns his roaring with baying, and gives him bite for bite. In such an encounter the dog may be worsted, but not till he has often severely galled and wounded the lion. The lion is however at times worsted by the Indian dog and killed in the chase. If a dog once clutches a lion, he retains his hold so pertinaciously that if one should ever cut off its leg with a knife he will not let go, however severe may be the pain he suffers, till death supervening compels him.

There are even more brutal descriptions of such dog fights. The lion in these situations is clearly from a hunting park or a menagerie.

Strabo speaks of the occasion when the king washes his hair ceremonially and a festival is held and big presents are brought as a display of wealth. He also writes about the animals:

> And in the processions at the time of festivals many elephants are paraded, all adorned with gold and silver as also many four-horse chariots and ox-teams... And tame bisons, leopards and lions and numbers of variegated and sweet-voiced birds.

Arrian, writing in the second century CE, describes Alexander's campaigns in some detail. His section on India, the *Indika*, has a summary of what an earlier author, Megasthenes, wrote about Mauryan India. There is a lengthy description of trapping wild elephants and much on tigers, parrots, apes and snakes, as well as a fascinating account of what he calls 'gold-digging ants'. But there is no reference to a lion hunt. The maximum interest in Greek accounts of India was in elephants.

According to Pliny, the area bound by the Indus River and surrounded by a ring of mountains and deserts lying between this river and the Yamuna, was infested with wild tigers. This would suggest modern-day Rajasthan and Gujarat. Pliny also refers to lions with and without manes, and the latter are thought by some to be leopards. Clearly, there is some confusion in identifying animals. Aelian, writing in the early centuries CE, believed that large lions are found in India, since it is 'an excellent mother' to so many other animals. He then remarks on how India's lions look almost black, 'the bristly hair of their mane stands erect, and their very aspect strikes the soul with terror and dismay.' Their malleability to being trained when captured young fascinated him, especially their being led by leashes to participate in 'hunting young deer and stags, and boars and buffaloes and wild asses' as they were known for their 'keen scent'.

This would again probably point to lions in the hunting parks as it would be more difficult to capture lion cubs from a pride in the wild, except by first killing the mother. Or is there some confusion here with the cheetah which was trained for this purpose if it was present in India at this early period? The more frequent references are to hunting dogs. And references to the cheetah are hard to come by, the more recognizable one being as late as the twelfth century CE.

THE CHEETAH

There is considerable uncertainty as to whether references to an animal that might be a cheetah are in fact to this animal or to the leopard or the hunting dog. The confusion is easily understood as the three animals have a certain resemblance and the characteristic differences are not mentioned by those who refer to them. Cheetahs, although belonging to the cat family, were sometimes thought of as dogs since they could be trained as royal dog-like pets. There is some confusion about animals of the chase and a lack of clarity in differentiating between the cheetah, the leopard and the hunting dog. Even trained cheetahs are seldom associated with hunting big game and are more frequently linked to hunting smaller, gentler species such as antelopes.

As compared to the graphic evidence of the cheetah being brought as tribute from the south and the north of Egypt to the capital, the data from India suggests the absence of the cheetah at an early date. Being brought as tribute implies that the cheetah was tamed in Egypt and probably used in hunts by the mid-second millennium BCE. Greek and Latin authors seem not to have been so familiar with the cheetah. One possibility that is discussed is the depiction of animals on a dish of the second century CE, where one animal may possibly be a cheetah.

References in Sanskrit texts are also not very clear and terms such as *dvipi* and *chitraka* (spotted), can refer to leopards and panthers as well. The first is used in earlier texts and the second in later works. The cheetah's environment is similar to that of the lion and, therefore, it would not be found in every part of the subcontinent even if it was known to the region earlier. One is tempted to suggest that the cheetah may have been a late import. That its habitat in Africa is the same as that of the African lion makes the connection more intriguing. Its presence can only be ascertained by reading descriptions of how it was trained and used.

The *Manasollasa*, written in the twelfth century CE, possibly has a reference to this. Many ways of hunting deer are listed. One of these is suggestive of the use of cheetahs as known from later times, although the identity of the animal is not certain. The animal is referred to variously, the description of the spots applying to other animals as well. The most striking characteristic of the cheetah which (apart from other features), differentiates it from leopards, tigers, panthers and the like, is the black teardrop line from the eye to the jaw. This is not mentioned in any of the descriptions; it is also omitted in the Chinese descriptions of the leopard-like hunting dogs.

In the course of giving instructions on how to train the animal to hunt, the *Manasollasa* uses various terms for the animal—*chitraka*, *vyala* (a general term for a variety of animals from elephants to serpents), *vyagrha*, *dvipi*. The description of the technique for making a kill is similar to that of the training given to cheetahs elsewhere. Even if the *chitraka* was referring to the cheetah, the training could be a relatively recent acquisition since it does not occur in earlier texts. It might be worth keeping in mind that the *Manasollasa* was written in the western Deccan at the Chalukya court, and by this time there were close trading contacts with Arabia and possibly East Africa too. Settlements of Arab communities were dotted along the west coast of India. The areas overseas, to which the trade was directed, were where the cheetah and the lion were familiar.

Hunting Parks

In lands where lions roamed wild in their natural habitat of grasslands, savannah forests and thickets, as in Africa and Mesopotamia, there were some obvious associations between the lion and big game that became significant features fairly early in the cultures of these lands. One of these was the establishment of hunting parks for the royal hunt in which the killing of the lion had priority in any claim to being an ace huntsman. Another was the rapidity with which the lion became the icon of the heroic, and following from that, of royalty and majesty. And then there was the amalgamation of the lion with a deity in the early pantheon of the culture.

All three of these features are present in Indian society but curiously occur rather late in its history. The emperor Ashoka writes in one of his edicts that his predecessors went on hunting expeditions which he stopped and converted into tours of his domains which he used as occasions to propagate ethics and the law. This pinpoints the fact that the hunt was a surrogate inspection of the kingdom when it went into areas outside the hunting parks. Ashoka probably disbanded the hunting parks as part of his policy of non-violence. Buddhists believed that those who hunted were reborn to a gruesome life and hunting would, in any case, have been contrary to the ethic that Ashoka endorsed.

References to hunting parks are not easily discernible in the early texts, but they seem to be obliquely mentioned in the *Arthashastra* of Kautilya, a text on political economy. Originally written in the fourth century BCE, it was revised and enlarged much later in the third century CE. This is the version we have now. As a text, it drew on the author's observations of kingship and its necessary institutions, and the hunting park was an institution associated with monarchy.

One of the chapters of the *Arthashastra*, while referring to the superintendent of forest produce, lists as produce to be collected by officers of the state, the body parts of animals, which included virtually everything. The animals mentioned include the lion, the tiger and the elephant. The forest here seems not to refer to the natural wilderness but to more limited forest areas. The term used is *dravya vana*, literally the wealth-producing forest, and it is further mentioned that the wealth produced should be processed, presumably for sale.

In another chapter, there is a description of what might be called a kind of royal zoo, the *mriga vana* or animal park, where a variety of animals were enclosed in a large area. It was located on wasteland that could not be cultivated and was presumably a planted forest.

> ...An animal park for the king's recreation should be laid out with a single entrance, protected by a moat...and stocked with tamed deer and other animals, and containing wild

animals with their claws and teeth removed, and male and female elephants and cubs, useful for hunting.

▪

The viciousness of this manner of disarming predatory animals to accommodate royal pleasures is horrendous, and competes in brutality with the spectacles in Rome where animals were pitted against one another or against a gladiator. This was clearly a reserved animal park but not used for hunting. The animals were doubtless also brought from another hunting or reserved park and were already under captivity, else it would have been difficult to remove their teeth and claws. One wonders, of course, about how many animals survived this treatment.

Adjoining this park, we are told, there should be another animal park where animals would not be 'disarmed' as it were, but be protected. This again would not qualify as a hunting park unless its being used as such is mentioned. Prime importance is given to the establishing of elephant parks, *hasti vana*, on the borders of the kingdom, where elephants were not killed but caught. However, this was a cordoned-off area of natural forest and the habitat of elephants from where some were trapped. Elephants caught here were tamed and trained for warfare and other heavy-duty work. The elephant was the actual royal animal on which the king rode, both in battle and in processions.

Another demarcated area of forest is described without specific reference to predators, but listing deer, animals, birds and fish. This was an enormous area, again isolated from the rest of the forest. Its location was not too far from the settlements. It is referred to as the *abhaya vana*, literally 'the forest where there is no fear', presumably a reserved forest. No mention is made of the king hunting in this forest but the clue may lie in a prohibition where severe punishment is meted out to anyone caught hunting on its premises. Some of these parks were under the supervision of the officer in charge of animal slaughter, and seem to have also been the source of the meat supply, governed by various rules of procedure. None of these are classic hunting parks of the West Asian variety, but may well have been of an alternate kind although serving the same function.

These descriptions of hunting parks and restricted forests are in strong contrast to the edicts of the emperor Ashoka, who orders the protection of animals both domesticated and in the wild. In one edict, he lists the animals that are not to be killed on particular days. It is a little unclear as to why these particular animals were listed since most are not those likely to be killed ordinarily. In the second century CE, Aelian relates that Indian kings had large royal parks more grandiose than those of Persia at Susa and Ecbatana, and that these were provided with man-made lakes, trees, shrubs and of course birds, fish and animals, many of which were imported from abroad.

The concept of the forest in India was neither singular nor static since the relationship to the forest changed over the centuries. Initially there was a dichotomy between the *aranya,* the wilderness, and the *grama*, the settlement. The *grama* was the known and the familiar, whereas the *aranya* was the dark unknown, the habitat of demons and thought of as a distant place hosting wild plants and wild animals. Gradually, the dichotomy became a little blurred. Cultivated fields edged into the precincts of the forest where hermitages had already encroached and the forest became more familiar. The terms used for the dichotomy were now the *kshetra*, the fields, the *vana*, the not-so-dense forest, and the *atavi*, the forest through which the inhabitants roamed.

The *vana* was where kings went hunting, where ascetics found a refuge and where those sent into exile had to while away their time. The forest dwellers were seen either as primitive humans or as demonic *rakshasas*. This was the forest featured in the *Ramayana* and the *Mahabharata.* Gradually with greater familiarity, the forest came to be seen as the romantic setting for poetic fantasy, such as in Kalisada's play on Shakuntala. Nevertheless, familiarity with the forest kept shifting with some part of it being better known than others. This is evident from the famous biography of King Harshavardhana, the *Harshacharita*, by Banabhatta. In the fringes of the forest there were scattered settlements, not of cultivators but of those who gathered the natural produce of the forest—beeswax, honey, medicinal plants, bark and such like—which were taken

Stylized lions on a pillar capital of the Kushan period of the first century CE. From the Mathura region, the sculptor may have been less familiar with lions. The inscription records royal patronage to a Buddhist place of worship as well as the genealogy of the local Shaka satrap, thus in a way linking the lion to royalty.

to the markets of the regular villages outside the forest. This was the median area where the natural forest had been tranquilized to enable the gathering of its produce and other activities. It was also the location of the hunting parks— the various *vanas* listed in the *Arthashastra*. Deeper into the forest was where its habitual inhabitants, such as the Shabaras, lived. They knew every leaf of the forest, moved through it like black clouds, and their chief had the predictable name of Vyaghraketu, 'the mark of the tiger'. What was inaccessible to others was accessible to them.

The relationship between the settlement and the deep forest was often one of conflict, with the former constantly encroaching upon the latter, even to the point of devastating it. The *khandava vana* was entirely destroyed in a horrendous conflagration, despite the fear and cry of the animals; this was done to make space for the Pandava settlement of Indraprastha. Ashoka threatens the *atavikas*, and the *Arthashastra* cautions against their presence since they can be destructive when resisting encroachment. Inscriptions of the late first millennium CE—when new kingdoms were mushrooming all over and looking for economic resources—refer to kings and ministers putting down the Bhillas, Pulindas, Shabaras, Nishads and other forest dwellers.

The forest, therefore, had a complex ambience hosting varied human relationships. The animals of the forests were those of the *vana,* for little is said of what inhabited the *aranya*. Its density may suggest that it was more likely the habitat of the tiger and less likely to have been the habitat of the lion.

▪

In the centuries CE, there is some royal association with lions. A gold coin of the Gupta king, Chandragupta II, depicts him as standing beside a slain lion. The legend refers to *simha-*

parakrama, the lion-conqueror. A similar coin was issued by a successor, Kumara Gupta. Was the killing of the lion by now a necessary qualification to claims of exalted kingship? But this is not a frequent motif. What is noticeable is that when referring to hunting, Indian kings do not mention the numbers of lions (or other predators) killed, as was the proud boast of Egyptian and Assyrian kings, and as was to become fashionable in later centuries in India. Was there a paucity of lions even in the hunting parks?

Kalidasa, in one of the opening verses of the *Kumarasambhava,* speaks of the Kiratas, the hunters of the northern mountains, recognizing the trail of lions that have killed elephants, not by their blood-smeared footprints in the snow, but by the pearls produced by the elephants that drop from between the lions' claws. Were there lions in the snow-covered mountains or was this a poetic mirage?

The lion capitals on columns continued to be an architectural feature well into the Gupta period of the mid-first millennium CE. It has been argued that the one from Udayagiri may have been a Mauryan original which was re-cut to suit Gupta aesthetics. It is a fairly common image on pillars such as the ones at Tigowa, Eran and Sanchi, which in form recall the Ashokan, but render the animal even less realistically. Had it by now become an idiom?

A relief of a lion and bull in combat from the palace of Darius at Persepolis dating to the sixth century BCE.

The Lion in Ritual and Belief

The lion as the metaphor of superiority and prowess was widely accepted throughout the ancient world and India was no exception. Mythology linked to the lion also travelled. The legend of Hercules and the Nemean lion, for instance, seems to be present in Gandhara in the early Christian Era. Even though not every part of the ancient world hosted the lion in the wild, it became a familiar motif as part of exotica, or circuses and menageries, and was adopted as a royal mascot. This is strikingly demonstrated in China where the lions came as gifts from other rulers, especially those from Central Asia, and the depiction of the lion became increasingly less leonine, and soon grew elaborate wings to become the *bixie*. The 'lion dance' that accompanied so many processions and rituals in China and Japan was amazingly popular, given that it was not an animal they knew closely. This does suggest that we should differentiate between animal representations based on familiarity with, or observance of, the animal and those that have their source in fantasy.

The association of the lion with deities in India, both male and female, again began relatively late and generally dates to the centuries CE. Vishnu incarnates himself as Narasimha, a man-lion, to exterminate evil and rescue his devotee. Mention is made of this *avatara* in the *Mahabharata*, but the longer myth is narrated in the *Shanti Parvan* which is an interpolation into the text of the centuries CE, and because of its being late is omitted in some editions. The theory of the *avataras* as incarnations of Vishnu is relatively late and although the incarnations are mentioned in scattered places, it only comes to be characteristic of Vaishnava belief from about the mid-first millennium CE. Of the non-human *avataras,* each had a major Purana dedicated to it giving the mythologies and the rituals required for the associated deity. However, Narasimha, not widely worshipped, had only an Upapurana, a minor Purana composed in about the sixth century CE. Although a deity, he behaved like a man-eating lion. The Narasimha image is found in many more temples of peninsular India than in North India and the images date to the mid-first millennium CE or later. Another myth mentions that the Narasimha incarnation was attacked by the Sharabha incarnation of Shiva—an articulation of sectarian hostility between the Vaishnavas and the Shaivas. This was a composite animal with eight legs, and in Shaiva temples it is predictably triumphant over the man-lion. The man-lion also takes shape

Relief in schist of Hercules with the Nemean lion from Gandhara, dating to the first century CE. It originates from the region to the west of the Indus and shows the difference in style and in the perception of the lion.

as the *purusha-mriga*, literally the man-animal, where the *mriga* usually looks vaguely like a lion but can be any similar animal, and the image can carry horns and wings. This, too, is found in Shaiva and Vaishnava temples. Another mythical animal seen in temples was the leogryph, a partial lion.

Female figurines currently interpreted as goddesses are known from Harappan times, and as a substratum religion they have continuity and universality ever since. The worship of the goddess through the Shakta-Shakti sects became prominent among elite groups only in the first millennium CE—a case of the upward mobility of a popular cult. Adding the lion to the persona of the goddess was combining the power of fertility with prowess and majesty, but the link was a later association dating to when the lion entered the repertoire of the Puranic Hindu religion.

Sejant lion in schist from Gandhara of the second-third century CE. It is somewhat reminiscent of the Sarnath lions of the Ashokan capital.

A range of goddesses from West Asia and the eastern Mediterranean, such as Inana, Ishtar and Ashtoreth, were shown either seated on a lion throne or sitting on a lion or standing by a lion. Was the iconography of Durga a parallel iconography? Or is there a hint of similar figures of goddesses with lions from Central Asia—such as Cybele with her chariot drawn by lions—as some have suggested? And did the iconography originally link Durga to a tiger and later change in some instances to a lion when the lion became an animal of majesty? The goddess riding the tiger is dominant in Tibetan iconography. The association of the lion with religious belief is a relatively late phenomenon. Shakta temples of the centuries CE, dedicated to the goddess, sometimes have sculptured lions as guardians. Elsewhere, a lion is shown trampling a crouching elephant in the precincts of a temple. Whatever the representations may be of the goddess and the lion, the early West Asian forms predate the Indian by some centuries.

Durga is shown to be standing beside a lion or riding a lion in icons from the first millennium CE. The form is somewhat rare at the turn of the CE but becomes prominent a few centuries later. This was also the time when the Narasimha incarnation becomes established in mythology. Durga is not consistently with a lion since sometimes it is substituted by a tiger. As Shiva's consort, she is sometimes referred to as Durga Simhavahini (Durga whose vehicle is the lion) and is shown with a lion. Ambika as a revered goddess is generally shown seated on a portly lion, also seated, and often decorated with a circular mane of snail-like curls, as seen at Ellora. By comparison, the predecessor goddesses of the ancient world associated with the lion—Sekhmet, Ishtar, Inanna, Anahita, Cybele—go back many centuries earlier. Once the popularity of this link was established in India, the lion was associated with deities among sects, even in areas of the subcontinent where people may not have seen a live lion except as an exhibit.

▪

The symbolism of the lion enters the architectural structure of religious buildings in many shapes and forms. Buddhist stupas and rock-cut caves have leonine forms as guardian figures (Amaravati, Sanchi, Karla) or holding up the seat of the Buddha (Mathura). The Pallava dynasty of South India (late first millennium CE), seems to have appropriated the lion symbolism and used it as an architectural feature at Bhairavakoda (Nellore district) and Mahabalipuram. The lion is either found at the base of the pillar or on its capital. But in the repertoire of painting and sculpture of early times, there are few (if any) spectacular scenes of hunting the lion when compared to ancient West Asia, or for that matter, the paintings from the Mughal era.

In sculpture, the mane of the lion takes various shapes—as strands, as curls (often snail-like), or as ringlets. Sometimes the lion is hardly recognizable with its frilly mane and elongated head (Daksharamam in Kakinada and Simhachalam in Visakhapatnam). It is possible that there were fewer lions in the South Indian landscape. But in sculpture, the lion generally seems to become less and less leonine. The ultimate fantasy was the *kirtimukha* (the face of glory), generally located over door-jambs in various temples, that converted the lion into a grotesque mask and, therefore, evoking its role as protector and guardian. One is left wondering how many of the sculptors—who sculpted these forms in various parts of the Indian subcontinent in early times—had actually seen a lion, or whether it had become entirely a matter of reproducing a feature of temple architecture?

Lions are not highlighted as the most magnificent animals of the forest and wilderness. Nor is there a hero like Gilgamesh or Hercules in Indian legends whose singularity lay in his overcoming the lion and thereby becoming iconic. And tigers and lions in their natural form are rarely represented as images of worship. Lions are prominent in fables and stories about animals, such as the *Panchatantra* or the *Hitopadesha*, as an animal whose authority was not questioned but who was not necessarily the wisest. In one well-known story, the lion was fooled by a hare. For wisdom, the choice lay more often with the elephant, an animal generally represented realistically in religious iconography, other than in the image of Ganesha.

In Europe, Leo became a personal name or name assumed officially by popes, kings and the like. In India, too, the word for a lion, *simha/sinha*, came to be used as a name, but seldom as a personal name and more as an identifier. Moreover, the adoption of this identifier by a community came to be used only in the second millennium CE. Dynasties may not have traced themselves back to a lion ancestry but by the early second millennium CE, some clans began to take the lion nomenclature by adding the name of Singh (from *simha*) to their names. This became common among Rajput clans and slowly spread to other parts of North India but was limited to particular clans and upper castes. The more widespread use of the name was a claim to status and came to be associated with land-owning families aspiring to be recognized as the kshatriya aristocracy, irrespective of what might have been their actual social origins. Needless to say, the kshatriyas of earlier times mentioned in the *Mahabharata*, the *Ramayana* and the Puranas had not used this name as a caste identity.

Those that were clearly not Rajput in origin, such as the royal family of the kingdom of Chamba in the Himalaya, nevertheless took on the name as a mark of status. Modern historians often explain this as an act of solidarity with the Rajputs opposing the Mughals, but this argument can hardly hold, considering that by the seventeenth century, the Mughals themselves had close blood ties in their kinship with the more eminent Rajput ruling families. The desire to be linked to the Rajputs, therefore, could have been due to the wish for supra-royalty connections rather than in opposition to them.

A still further adoption of the name came with the growth of Sikhism in Punjab some centuries later. As a large number of people converted to the faith and as militancy was inducted into their activities, 'Singh' was added on to the name and became a form of religious identification. This was then explained in popular usage as asserting power, as does the lion. The universal use of the identifier among Sikhs was also an attempt to blur caste distinctions, but such distinctions remained and were sometimes introduced as the equivalent of a surname. The erstwhile identifier was now replaced by a religious label. There is, of course, no kinship link between all those that use this identifier since it was picked up by a variety of North Indian communities late in Indian history, by when the symbolism of the lion was well established. Interestingly, this was the period of the Sultanates and the Mughals, when hunting parks became a more marked feature of the landscape. Actual lions, few in number, were the prized animals that were brutally hunted down in these parks. No attempt was made by those who carried the name 'Singh' to object to this form of hunting. The Rajput elite participated in it, fully and gladly.

Where Did the Lion Come from If it Was an Exotic Alien?

Let me go back to the initial question of whether the lion was widespread in the Indian forests or whether it was brought to India for various purposes. Did it subsequently go feral in some areas? If it was introduced into India, then where could it have come from?

The journey from Africa would have involved a long voyage

across the Arabian Sea, caged in the hold of a ship, accompanied by those who could handle the animals—a task of immense difficulty but not impossible. The handlers would have been those who were familiar with the animal from Africa. But the evidence of such an import of animals together with their handlers, either as a gift package or as part of trade, is not forthcoming for the first millennium BCE. The presence of the lion in India around the CE is noted by Greek and Latin authors; it is possible that advances made in ship building and navigation and the lucrative maritime trade in importing horses from the West may have made it easier to transport the exotic animals. Another possibility would be bringing lions from West Asia, given the traffic in sizeable cargo along the routes that linked North India to the eastern Mediterranean. Nevertheless, such consignments would tend to be small although in aggregate they would have added up.

A heavily stylized lion rearing over an elephant at the Surya Temple at Konark.

Perhaps a more likely source for lions could have been Balkh/Bactria which was in early contact with northwestern India. The Achaemenids hunted lions in the Oxus valley in the sixth century BCE, if not earlier, and the hunts continued into later times. Alexander is said to have hunted a lion north of the Oxus, an area where lions seem to have been common, either in the wild or in hunting parks. He is likely to have been taken to the same area for hunting lions as where the Achaemenids hunted. In contrast to this, there is no mention of Alexander hunting lions during his lengthy campaign in northern and western India. Was this because, having crossed from Balkh to Afghanistan and then into the southern part of the Hindu Kush and Pamirs, he was no longer in lion country? The association of Bactria with lions continued unabated. A second century CE plate from Ai-Khanoum in

Bactria shows the goddess Cybele riding a chariot drawn by two lions. Lion hunts are frequently depicted on a variety of objects whose provenance was Bactria and Iran, and this continued well into Sassanian times. The Chinese received the occasional gift of a lion from Samarkand.

Chinese Buddhist pilgrims coming to India via Central Asia in the fifth and sixth centuries CE, saw lions being sent as tribute from Balkh to the king of Gandhara. This was by now a well-travelled route between Northwest India and Central Asia. The other route discovered more recently was in the vicinity of the Karakorum Range and went via Gilgit and Chitral. When traced, this ancient route was found to be flanked with inscriptions, engravings and paintings of a Buddhist nature and dating to the first few centuries CE. Yet Mathura, in the western Ganges plain, provides us with rather cheerful and chubby winged lions.

▪

That Bactria, northeastern Iran, eastern Afghanistan and northwestern India constituted an area with constant and close contact is evident from its political and cultural history through many centuries. It often came under common political control—under the Achaemenids, partially with the Mauryas, and then again with the Indo-Bactrian Greeks, the Shakas, Kushanas, Sassanians and to some extent, the Shahis; not to mention subsequent dynasties associated with the expansion of Turkish, Afghan and Mughal polities. This meant that the proximity of Bactria to northwestern India continued for over a millennium and more. The function of Hellenistic settlements was to pursue trade connections as is evident from the scatter of Hellenistic cities in the area and what their excavations have revealed. Trade routes were well established and there were further links with the growth of the Silk Route connection from China to Byzantium via Central Asia. The horse entered India from Bactria together with the Indo-Aryan speakers in the second millennium BCE, and the trade in horses continued unabated by using all the viable routes. Bringing horses from Central Asia was a major commercial enterprise in which even Brahmins were prominent traders.

Admittedly, horses are easier to transport than lions but the profit motive surmounts all problems. Caged lions from Bactria could have been part of the extensive cargo that travelled these routes, given that there would have been a rich income on arrival.

Unfortunately, the texts that speak of trade between various parts of India and places to the West and the North do not mention the coming and going of large animals. The trade across the Arabian Sea was in plants (primarily pepper and spices) and in textiles and gemstones—all small objects in return for Roman coins and amphorae of wine and other such goods. However, animals such as the one-horned rhinoceros and tigers, together with peacocks and parrots, are mentioned as common to the Deccan. These animals find their way to the Mediterranean, together with pythons, as gifts. If such large animals could be transported, then so could the lion. But there are no references to lions, neither coming to India nor being sent out of India in these times.

Ships from West Asia came to ports along the west coast of India. These coastal areas were home to many settlements of Persian and Arab traders whose members were employed by local rulers, for which there are records from the eighth century CE onwards. These were the ports where horses were landed and from where they travelled to the hinterland in various parts of western and peninsula India. If this trade also brought lions and cheetahs from Africa, they would have come with handlers, as did the horses, and some may have been kept in areas such as western India, where the climate and environment suited the animals. The location of lions as given in Greek and Latin accounts is more frequently the western region of India. What is not certain in these accounts is whether Greek and Latin authors, when referring to lions in India, were referring to those from hunting parks or in the wild or both. That some were hunted and trapped would point to their being brought from hunting parks or menageries, particularly those described as tame lions paraded in processions.

But to return to the main question: was the lion indigenous to India from earliest times? Had the lion been viewed as an exotic animal from India, it would most likely have travelled west as part of the huge gift packages that were sent to the Roman emperors and others. Augustus received an embassy bringing tigers and pythons. Peacocks and parrots and such like accompanied delegates visiting Trajan. Constantine, said to be much revered by Indians, received the same. But lions seemed to come to Rome, not from India but from Africa.

The evidence does not unequivocally support the existence of the lion as a habitué of India since very ancient times. It gradually builds up to a presence in various forms of representation by the centuries CE, much later than in the West. In Africa, Mesopotamia and Bactria, lions were continually sighted going back to much earlier periods and were a part of the landscape. Sighting is possible, in part because of their habitat which is open grassland and bush, and because lion prides are large in number. Tigers by contrast live in thick forests, are more solitary and, therefore, comparatively difficult to spot.

Yet the presence of the lion in India is recorded, later of course than the initial civilization of the Harappan cities. Although there are passing references to the lion in the texts of the next millennium, it is not the single pre-eminent animal or the prize animal hunted in the wild. A comparison with the information from Egypt and West Asia highlights the difference in how the animal is viewed. The data from these areas, where lions were hunted both in the wild and in parks, is strikingly dissimilar. This, I think, is important in assessing the presence of the lion as a pre-eminent animal in India from earliest times.

To sum up, although hunting parks or something similar are known to have existed, the hunting of lions, whether in these parks or elsewhere, are not the prize hunts of the early period. The popularity of this activity seems to have developed much later. The question of the lion being indigenous, therefore, still requires investigation. It would seem that the lion was a latecomer to the pantheon of the great animals in India, being preceded by the tiger, the rhinoceros, the elephant and others. Nor did kings claim to have killed huge numbers of lions in the hunt. Perhaps the numbers of lions in India were few as compared to other

The Alexander Sarcophagus at the Royal Necropolis of Sidon in the Museum of Archeology, Istanbul (fourth century BCE).

kingdoms and hunting them may have often been confined to hunting parks.

Speaking impressionistically, it would seem that the lion arrived on the Indian landscape at the period when Mesopotamia, Iran, northwestern India and the Oxus Valley had close connections around the Achaemenid period. In a sense, this connection may have been reiterated when Alexander traversed these regions. If this impression reflects reality, then a familiarity with the lion in northern India may date to these times. Subsequently, the lion had the same symbolic role in India as it had in the empires to the west of India. I have tried to present the evidence as I perceive it but it requires more detailed investigation before a conclusive generalization can be made. What the evidence does, however, state quite definitively, is that the lion, whether indigenous or not, was a recognized metaphor of dignity, majesty and prowess in the many cultures of India.

The first representations of gifts and tributes of cheetahs go back nearly 5,000 years. This cheetah was being taken as a tribute to the king of Thebes in 1700 BCE.

Next page:
The tradition and rituals connected with the slaying of lions probably had their origins in Africa, where man and lion first met. Travelling through time, the Romans probably imported by sea nearly 1,000 lions from Africa between the first and third century CE for the gladiatorial games. The question is, what did the Indian kings import? The gold coins of Chandragupta (fourth century CE) show him slaying a lion. Were these lions imported? All we know is that the rulers of the world portrayed themselves as slayers of lions from times immemorial, irrespective of whether lions lived in their countries or not.

ANIMALS OF THE CHASE

by YUSUF ANSARI

The Austrian painter, Wenzel Peter's depiction of Genesis—the beginning of the world—has at its centre a pair of lions, a male and female. The presence of these big cats in the depiction of Adam and Eve in the Garden of Eden grants the lion a position of prominence that effectively supersedes time and geographical boundaries. Subliminally, human sensibility has accepted the existence of the lion as perfectly natural in the political, social and cultural depictions of a variety of civilizations and historical contexts, even in those where there is no evidence that it naturally existed. Historically then, the lion has acquired a cultural omnipresence like no other beast, real or imagined.

The cheetah's claim on human history is also very old. An illustration in the *Book of the Dead* of Ani (1250 BCE) from ancient Egypt portrays an animal astride Pharaonic hieroglyphs supporting the sun. At once, this whimsical creature, both lion and cheetah, combines two symbols of royalty and power. Scholars aver that 'the earliest evidence for cheetahs under human control' is to be seen in the famous Punt reliefs at Dayr el-Bahari, across the Nile from Thebes. These record, in word and picture, an exhibition sent during the reign of the pharaoh queen, Hatshepsut (1473-1458 BCE), to the land of Punt somewhere in the Horn of Africa, which brought back to Egypt animals such as 'panthers'. Two very distinct types are shown in these clearly-executed reliefs. The first, composed of two cheetahs on leashes, is labelled 'Panthers of the North'.

Potently, the journey of the lion and the cheetah is a part of the narrative of globalization from ancient times, a finely choreographed progression through human history that dispersed them far from their original habitats. This was especially true of lions that were prized because nothing symbolized the conquest of nature better than the conquest of its most feared terrestrial representative. Lion slaying became a monarchical monopoly and the preserve of kings, the rite itself attributing to them a noble and superhuman ability to destroy or control savage elements. In the previous chapter, the place of the lion in ancient record has been covered in detail, as has that of the cheetah as a hunting animal of trans-imperial elites. Here, we shall focus on the place of both species in India, from the earliest Islamic arrivals into the subcontinent in the eighth century to the late Mughal period. Momentarily, it would be useful to reconsider this book's central thesis.

At what stage does a migratory species earn the distinction or label of becoming a native one? Lions are 'native' to India today, but their arrival into India is clouded in conjecture, and subsequent studies on the evolution of the species are, at best, speculative. As we have seen, one premise that converts the provenance of the Indian lion from its mythological foundations into historical reality is the presence of game parks and hunting reserves throughout its former range, where the animal was enclosed, bred and stocked to be hunted and as material for diplomatic tribute and spectacle. Persian, Greek and Assyrian accounts all suggest that these animals were subsequently exported or found their way through trade and diplomatic channels into new ranges. This could be how the Asiatic lion (*Panthera leo persica*) arrived in India; 'naturalized' over time as the centuries passed, the lion was accepted as an organic, 'native' species of the country. Certainly, this holds more water than theories of the Indus drying up, allowing the lion to cross over from Persia during periods of drought.

The range of Hinduism and the boundaries of India (as we know it today) were not the same at any given period of history.

Ramesses II returns home after a victory with a cheetah by his side (early 1800-1843 BCE).

Hindu, or more properly, Vedic religious thought, was more active in the Persian-Sogdiana region at a particular period and may have been very familiar with the lion, before the lion familiarized itself with India. Likewise, Vedic literature, with its wide geographical and cultural ranges, played its own part in bringing the lion into the Indian cultural mainstream. This accounts for the presence of lions in India's cultural space despite the lack of physical evidence of the animal within India's geographical boundaries.

Although the lion and the cheetah were embedded in the art, culture and religions of India, are there any conclusions we can draw about the presence of these two animals on the ground through the historical record? Evidently, as we shall see, the range of both the lion and the cheetah in India corresponds with the existence of ancient and medieval equivalents of royal game parks (for hunting) and the locations of royal and feudal menageries across the country. Numerically and in terms of location, sightings or killings of these animals do not suggest they existed as organic, indigenous creatures of the wild, as has been supposed.

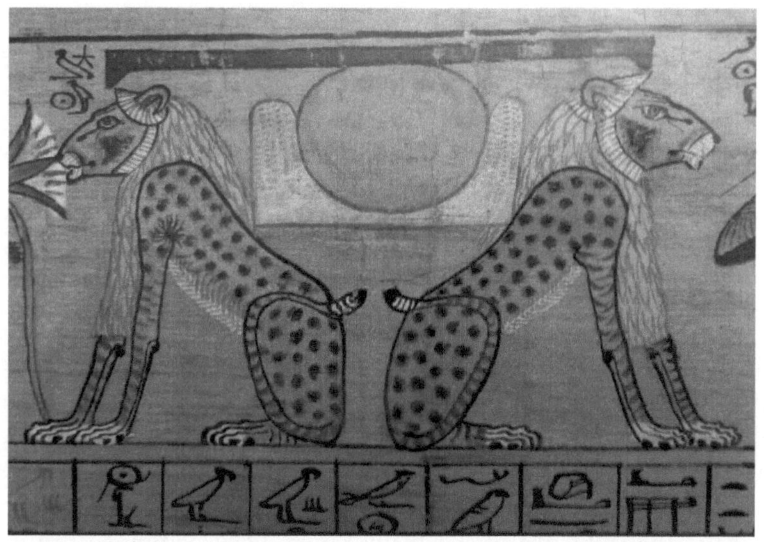

Papyrus of Ani—two lions representing yesterday and tomorrow (1295 BCE). In actuality, they are 'maned' cheetahs.

While we can only speculate about the provenance of these species, what is beyond doubt is that, for Islamic arrivals into the Indian subcontinent in the eighth century, 'India's lion' was not a discovery. The lion had a long history among the pre-Islamic Arabs who worshipped Yaghūth, the youthful man-lion war deity. Certain tribes regarded him as a protector and his idol is believed to have adorned the Ka'abah, in Mecca. He was a prominent enough deity to merit an expedition by Muslim forces soon after the conquest of Mecca, aimed at the destruction of his idol at Tai'f. With the triumph of Islam in Arabia, the lion was incorporated as a symbol of valour and bravery by Muslim chroniclers. The following reference illustrates some of the characteristics that defined the Abbasid court in Baghdad in the tenth century CE. Of particular interest is a passage describing the reception accorded to ambassadors from Byzantium in 917 CE:

> ...Then they were taken down corridors and passages to the wild animal enclosures... Here beasts were brought in from the garden and came right up close to the people, sniffing them and eating from their hands... Finally, they were taken

> to the lion house, where there were fifty lions to the left and fifty to the right, muzzled with chains, each held by its keeper.

Nor were cheetahs and their hunting skills discovered suddenly by Islamic arrivals into India. In fact, in India:

> ...hunting with cheetahs did not become popular among the ruling classes until the eleventh century and the custom may have been learnt from the Muslims.

The early thirteenth century Islamic rulers of Delhi, who were subsequently amalgamated into what we know as the Slave Dynasty, were an energetic bunch. Descendants of slaves serving at the court of Delhi's ever-changing sultans, these formidable individuals were initially picked out for their loyalty and valour to expand the influence of the Delhi court. They combined private austerity with a generous spirit and laid the foundations of a kingdom whose boundaries would serve to rouse the ambitions of later dynasts. The Qutb Minar is an enduring symbol of their resolve. In time, intermittent monarchs of this line fell into decadence, even debauchery, concentrating their time and energy inside the harem and its politics. The Khiljis who succeeded them expanded these borders, reaching as far as the Deccan in their quest to create a pan-Indian empire. The Tughlaqs and Lodis who followed the Khiljis ignored the latter's empire-building models to their peril. Nonetheless, this rich melange of sultans, flushed with the diverse ancestry of thousands of Central-Asian, Turkish, Persian and Afghan warriors all had one thing in common—their love of the shikar.

In the Islamic context, hunting game with 'animals of the chase' finds sanction in the Qur'an itself. Nothing less than divine ordinance allows the use of trained 'hunting animals'. The passage below is remarkable for the specific sanction it provides for hunting with trained beasts, a legitimacy that must have provided comfort to Muslim potentates throughout the Islamic world:

> They ask you, [O Muhammad], what has been made lawful for them. Say, 'Lawful for you are [all] good foods and [game

caught by] what you have trained of hunting animals which you train as Allah has taught you. So eat of what they catch for you, and mention the name of Allah upon it, and fear Allah.' Indeed, Allah is swift in account.

Cheetahs across the world were transported by horses to sites from where they were unleashed for the hunt. In India after the seventeenth century, bullock carts were frequently used to transport cheetahs.

Nearly every sultan who sat on the throne of Delhi found merit in hunting and the chase. Akbar's reign (1556-1605) would see the cheetah assume a position at the Mughal court which even noblemen would have envied. Historical accounts of Akbar's apparent obsession with collecting cheetahs overlook the fact that cheetahs were entangled with Islamic dynasties long before the Mughals (and Akbar in particular), made them inseparable royal companions. Nizami states that Qutb-ud-din Mubarak, the third Khilji monarch (1316-1320), possessed 'two to three thousand deer hunting panthers', a figure which, if true, makes Akbar's one thousand look like an amateur collection! For our purpose, this narrative only buttresses the point that Akbar inherited an Indo-Islamic culture that was well versed with the cheetah. Even the most austere and forbidding of the sultans of Delhi, Ghiyas-ud-din Balban (1266-1287), was a legendary hunter. Certainly, he must have been accomplished enough to merit mention in the courts of Baghdad, as Barni records:

> He took great pleasure in hunting, and followed it with much zest during the winter. By his orders the country for twenty kos round Dehli (present day Delhi) was preserved, and no one was allowed to take game. He used to go out in the morning, and always returned at night, even if it were midnight. A thousand horsemen belonging to the palace guard, each man of whom was acquainted with his

person, accompanied him; besides a thousand old and trusty footmen and archers. Reports of the hunting expeditions of the Sultan were carried to Hulaku in Baghdad, and he said, 'Balban is a shrewd ruler and has had much experience in government. He goes out apparently to hunt but really to exercise his men and horses, so that they may not be wanting when times of danger and war arrive.'

Even as records begin to emerge from the mists of myth into the clearer air of history, the cheetah is still elusive. Nonetheless, there is a definite movement towards a more precise, elaborate and detailed record keeping from the fourteenth century onwards. Indeed, the evidence before this is scant and at best speculative and even Divyabhanusinh concedes, 'It may well be that the cheetah was not as common in ancient India as it must have become in medieval times, along with the blackbuck, its preferred prey, due to large forest areas being cleared to make way for open grasslands.' Some contemporary accounts contradict this view and argue the reverse: that areas were deliberately 'designed' as wastelands (or wilderness) to encourage the proliferation of prey and predator species for the hunt and diplomatic stockpiling. Forest management under the various dynasties of Delhi from the twelfth century onwards happened for a variety of reasons. Just like their predecessors all over the world, the fourteenth century Tughlaq monarchs maintained captive animals in large hunting reserves. Evidence suggests this was part of Firoz Shah Tughlaq's policy of land management:

> The chase of the deer was carried on principally in the neighbourhood of Badaun and Anwala, where these animals were found in great numbers. This district was [a] waste[land], but well furnished with water and grass. No other such waste was to be found near Delhi. Orders were given for its being retained waste for hunting purposes, otherwise it would quickly have become peopled and cultivated under the prosperous and fostering government of Firoz. If a lion, tiger, or wolf was surrounded, the Sultan used to kill it first, and then pursue the other animals.

The painting depicts a massive Mughal 'canned' hunt whereby the game is driven towards the centre of a ten-mile circular area so that the emperor and his entourage could hunt and kill the animals. Also seen are the walls of the enclosure with an army of men lined up outside (1590-1595 CE).

Firoz Shah Tughlaq's achievements at construction were matched by his collection of animals. Fawning chroniclers are barely able to contain their excitement when they record at length, the exotica accumulated by the sultan in his menageries and game parks. Clearly, not all of these were acquired without difficulty, and nothing less than divine clemency is cited for some of the more remarkable beasts in these congeries:

> Since the Monarch, the Elect of the Merciful Lord God, was interested in hunting matters, during his reign he took game of various sorts and devoted much effort to this, taking innumerable game of every kind. He had obtained beasts of prey of every variety.

Although the chronicler does not mention the method by which the sultan 'obtained' such a diverse collection of prey, he is extremely specific about the species of predators Firoz Shah collected:

> Of the single class of cheetahs and leopards he gathered so many that it is beyond computation, and of the class of caracal, so many that it cannot be stated in writing, and of the class of dogs so many that it surpassed all comparisons.

Here the writer pauses, as though to take in the enormity of recording the following sentence before proceeding: 'Indeed, during his reign, by the inspiration of the Lord God, Sultan Firoz Shah acquired a number of lions.'

The acquisition of a lion appears to astonish even the court chronicler, and can, therefore, be regarded as a seminal achievement. Rather than indicate that lions were to be readily found and hoarded, the Tughlaq chronicler suggests a divinely inspired breakthrough that helped Firoz Shah to achieve what is clearly seen to be an extremely difficult enterprise. In any case, the narrative undermines the notion that lions commonly ranged all over North India, even in the fourteenth century, and stresses the acute rarity of the animal.

Firoz Shah's love of the chase was legendary, and even survived the admonitions of saints the sultan had great reverence

Akbar hunting with cheetahs at night. The Mughal penchant for hunting in their private game parks was at par with the extravagance of their courts.

for. His chronicler, Afif, recounts an incident which throws some light on this:

> The Sultán...marched in great state from Karoda... After several stages he arrived at Hánsí, where he went to wait upon the Shaikhu-l Islám Shaikh Kutbu-d dín. The Shaikh said to him, 'I have heard it said that you are addicted to wine; but if Sultáns and the heads of religion give themselves up to wine-bibbing, the wants of the poor and needy will get little attention.' The Sultán thereupon said that he would drink no more. After this the Shaikh said that he had been informed that the Sultán was passionately fond of hunting; but hunting was a source of great trouble and distress to the world, and could not be approved. To kill any animal without necessity was wrong, and hunting ought not to be prosecuted farther than was necessary to supply the wants of man—all beyond this was reprehensible. The Sultan, in reverence of the Shaikh, promised to abstain from hunting.

Perhaps the story is an apocryphal one, for it did little to put a stop to Firoz Shah's love of the chase. His passion for hunting would lead him to design the 'Khushq-i-Shikar', a hunting palace within reach of his capital at Delhi. In fact, so engaged was the sultan with the proper maintenance of his hunting grounds that he diverted water from the Jamuna through a system of canals to irrigate his game reserves in Haryana. Thanks to Firoz Shah's perseverance, these reserves thrived. Succeeding dynasts of Delhi would continue to hunt in these wilds of Haryana for centuries to come. The District Gazetteer of Karnal recounts: 'In old times lions and tigers were not uncommon in the tract. The Nardak was a favorite spot for the old Emperors to hunt lion in.'

Firoz Shah Tughlaq had literally created the grounds upon which his nemesis, the Timurid dynasty, would reign, prosper and pleasure themselves. To the *paradeiso* of the Assyrians, the *veranio* of the Greeks and the Mauryan *vana* had been added the Indo-Islamic *shikargah*. In the long history of the royal hunt, its rituals and symbolism, both the lion and the cheetah were about to enter a new age in their own enigmatic history in India.

THE SHAHENSHAH'S SHIKAR

by YUSUF ANSARI

Adham Khan pays homage to Akbar at Sarangpur. Carrying tributes and gifts of cheetahs were common and keeping cheetahs as pets was fashion.

Next page:
The Assyrian lion hunts followed a model that must have been time-tested. Lions were received as gifts, bred in menageries and released into private hunting parks for slaughter. It was the greatest show on Earth and a public display of power over the people. These reliefs from 645 BCE of Ashurbanipal killing a lion in 883-859 BCE are themes that traversed the world up until the twentieth century in India.

> The special powers and properties attributed to successful hunters derive, in no small measure, from the special powers and properties attributed to the animals they vanquish. Fabulous beasts can only be slain by fabulous humans.

The Mughals were perhaps the most fabulous of all the dynasties to rule a larger contiguous land mass of India than any preceding or following them, and for a longer period. In the seventeenth century, the Mughal emperor presided over an empire of over 140 million subjects and had larger revenues than any other monarch on the planet, the Ottomans and Safavids included. The chronicles of the Mughal empire describe the fauna (and flora) of Hindustan more vividly and in greater detail than any records before them. Not surprisingly then, both the lion and the cheetah find a focus of attention they had hitherto not received in the accounts of lives and times of many of the Mughal emperors and their nobility.

Their Timurid ancestry formed the cornerstone of the Mughals' identity. Babur's invasion of India sought legitimacy—if any was needed—through his direct descent from Timur, whose empire had included northwestern India, even Delhi. Genealogical connections alone, however, were an inadequate reference for the Mughals to claim a Timurid inheritance, and by implication, the larger Persian heritage which was a part of

Timur's imperial creed. Emulation of his deeds, the adaptation of his symbols as their own and their constant preoccupation with establishing this primordial link with Timur affected the evolution of Mughal habits in India. In this elaborate programme, the lion hunt was an ancillary, albeit an important one as ritual, in the formalization of a Timurid cult. One of the instances we have of Timur killing lions occurred, unsurprisingly, near Balkh, in the Bactria-Sogdiana region.

> During the march '(near Balkh)' two lions made their appearance, one of them a male, the other a female. I resolved to kill them myself, and having shot them both with arrows, I considered this circumstance as a lucky omen.

Mughals would continue to consider the killing of a lion auspicious, its escape from a hunt, a bad omen.

Although the *Baburnama* has detailed accounts on the wildlife of Hindustan, Babur himself does not give details about either cheetah or lion in this fascinating memoir. Humayun, too, is silent about any encounters he may have had with these two beasts. During his exile in Persia that was enforced by the Afghan interlude when Sher Shah Suri had replaced him as the emperor of Hindustan, Humayun encountered Safavid courtly rituals regularly. Narratives of his life leave us with few details of the hunts he participated in, but knowing what we do about Persian court etiquette, it is reasonable to conclude that he accompanied the Shah of Persia on his hunting excursions. The Persians were relentless hunters and breeders of lions and it is probable that later Mughal menageries as well as captive lion populations in their hunting reserves had a Persian provenance. Until as late as the nineteenth century, accounts from Persia suggest a continuation of the ancient ritualized hunts, where lions were stocked and bred in what one may call 'organic studs', so emperors, shahs and sultans could project their own royal power through a mastery over these beasts. Visitors to Persia recall:

> ...the pleasantly-situated palace of Dawshantepe (which means in Turkish 'Hare-hill'), where the Shah often goes

> to pursue the chase, to which he is passionately devoted. This palace, of dazzling whiteness, stands on an eminence to the north-east of the town, and forms a very conspicuous feature in the landscape. Besides the palace on the hill there is another in a garden on its southern side, attached to which is a small menagerie belonging to the Shah. This collection of animals is not very extensive, but includes fine specimens of the Persian lion (shir), whose most famous haunt is in the forests of Dasht-i-Arjin, between Shiraz and Bushire, as well as a few tigers (babr), leopards (palang), and baboons (shangal).

There are certainly visual renditions of Humayun hunting rhinoceros near Kabul, but we have no accounts of him hunting lions or coursing with cheetahs. Nor are there records of either Babur or Humayun encountering cheetahs or lions in the wild, though tigers are mentioned. Sidi Ali Reis, the 'marooned' Ottoman admiral who travelled back to Turkey and passed through Humayun's kingdom between 1555 and 1556, wrote an insightful account of his journey. His is, perhaps, the only first-hand record that mentions lions during the reign of Humayun, though this occurs only after he leaves the newly reclaimed Mughal realms behind. Cutting across Hindustan diagonally from Gujarat to Delhi, he notes:

> We were struck with the enormous size of the crocodiles sporting in the river, as also with the numbers of tigers on the banks.

Ali Reis spent a significant period of time traversing Gujarat, particularly 'Djoona' or Junagarh, Kathiawar and many other acclaimed 'lion habitats'. His observer's eye mentions catching gazelles from the wild and using them for hunting as well as spotting wild buffaloes. In none of his descriptions of India does he mention lions. Not until he gets to Balkh and the Uzbek territories outside of India does the admiral come across his first lions, and by his accounts, they appeared to be quite a handful:

> We were obliged to fight with the lions day and night, and no man dared to go alone for water.

That is about as close as we get to lions in the reign of Humayun. It may, therefore, be reasonable to suppose that in the time of Humayun and his father, Babur, wild lions could only be found outside their newly acquired territories in Hindustan. Their own memoirs, and in the case of Humayun, the recollections of his sister, Princess Gulbadan, and his valet, Jauhar, are silent about the cheetah and the lion.

▪

Like Alexander the Great before him, Timur launched his campaign against India backed by the wealth of Central Asia. Yasi (the Turkestan of later annals), which Timur passed through before his descent into India, '…was a thriving town on the caravan route, its markets home to merchants trading tiger skins, gold and silver from Persia, porcelain from China, astrakhan, glassware, Siberian deer, lynx and the ubiquitous silks.'

Given its location, Turkestan would continue to remain a centre of trade in exotica, material and animals for centuries to come and may well have been a hub for the dispersal of captive lions and cheetahs between China, Persia, India and even Tsarist Russia. From Yasi, Timur's forces—part caravan but largely an army of conquest—meandered towards Kabul stocked with loot from his conquest of Persia. Although on a military campaign, he was traversing long established trade routes—which doubled up as avenues for invasion—and which regularly saw the transportation of exotic beasts, including that of *janwar-i-shikar*, or animals of the chase, into Afghanistan. This veritable supply line continued to be a conduit for later Mughal reigns.

During Timur's stay in Kabul, Shaykh Nur ad-din, a local notable and vassal, presented him with all the wealth that had been collected as a result of a five-year plunder of Persia that had concluded in 1396. 'He brought with him an immense treasure,' wrote Yazdi, 'with abundance of jewels of inestimable price; likewise animals proper for the chase, and birds of prey; leopards, gold money, belts enriched with precious stones, vests woven with gold…'

In a near replication of Alexander's reception in Babylon

(described earlier in the book), Timur was also serenaded with, among other things, hunting animals. The leopards mentioned here may well be cheetahs, as 'animals proper for the chase' would suggest. This vast treasure wound its way down the Hindu Kush, into Sindh and emptied out on to the plains of Delhi, which Timur pulverized in a shocking campaign in 1398, destroying Tughlaq power forever. Could some of the animals accompanying Timur have found their way into the hunting reserves so painstakingly created by Firoz Shah more than a century previously? An avid hunter, it is unlikely that Timur did not indulge in this pastime during the period he spent in the conquest of Hindustan and it is possible that some of the animals did not accompany him on his long journey back to Samarkhand. He certainly took Indian war elephants to Samarkhand with him, contributing once more to the flow of species across empires and geographical boundaries.

■

Although lions seem to have been passed over in Humayun's reign, it was during his invasion of India to reclaim the Mughal empire in 1555 that we get our first full account of a cheetah in Mughal times, albeit a captive one. The cheetah is described in the *Akbarnama*, the chronicle of the reign of Akbar, Humayun's son. In it, the author, Sheikh Abu'l-Fazl, describes the wonderment with which Akbar regarded the cheetah, which was one of the tributes presented to Akbar after his father defeated the Afghans in the Battle of Machhiwara in 1555.

> This was the first time that he conceived an inclination for hunting with the cheetah, and the first place where he saw the sport. For Wali Beg, the father of the Khan Jahan, presented as peshkash a cheetah which had come into his hands from the Afghans at the battle of Machhiwara and was called Fatehbaz (the Gamester of Victory). The farseeing who were in his entourage made conjectures about seeing this strange form. The keeper of this cheetah was called Dundu and on account of his good qualities, he received the title of Fateh Khan.

From this account, it would appear that for Akbar and his followers, this was a first encounter with the cheetah. The proposition is a difficult one to believe, as many of Humayun's aides and confidantes had shared his exile in Persia and would likely have come across the cheetah there, even if they had never seen one in India before. For Akbar though, this may well have been his first *yuz palang*, or hunting leopard. So far, the young prince had coursed with dogs, and indulged in falconry during his childhood in Kabul. This is also Akbar's first recorded receipt of a tribute, and perhaps why he developed a fascination, veering on obsession, with collecting cheetahs all through his lifetime. Nonetheless, the excitement and amazement in the author's voice at the 'strange' sight of the animal is a reflection of Akbar's reaction, which would have been faithfully reported by Abu'l-Fazl as the court chronicler. What is surprising is the absence of any cheetahs being recorded all along the plains and grasslands that Humayun and Akbar covered through the Sindh and Punjab, before their advance culminated in the Battle of Machhiwara. After all, these were prime cheetah landscapes.

Akbar, the third Mughal emperor and considered by many as the greatest of all among that heady dynasty, popularized the cheetah and granted it historical immortality in India. The cheetah, rather than the lion, therefore becomes the focus of our examination when we explore Akbar's age for clues to the origins of these big cats in the Mughal narrative.

Since his introduction to the cheetah in 1555, Akbar developed an obsession for coursing with the animal. The young prince happened to be hunting in the Punjab plains during a lull in his campaign to subdue Afghan rebels when he received news of his father Humayun's death. Crowned king at the age of thirteen, Akbar found respite and diversion in this sport in his darkest days, until it became a lifelong habit.

However, the accounts of Akbar's cheetahs are nothing less than a litany of fantastical notions that can be affirmed by neither historical evidence nor scientific plausibility.

The first hypothesis which needs rebuttal is the supposition that the emperor amassed 1,000 cheetahs in his reign and the assumption that these were all domesticated or stayed about his person. This claim has always been viewed with suspicion by serious historians but needs a conclusive burial. The contention is frivolous, to say the least, for the following reasons: No royal establishment in Akbar's reign, or those of his descendants, was built to house 1,000 cheetahs. None of the cities he based himself in—Fatehpur Sikri, Agra, Delhi or Lahore—possessed kennels that could house so many pet cheetahs. Moreover, feeding these numbers of cheetahs was another matter. The *Ain-i-Akbari* is meticulous in its records of the management of the royal menageries and the *cuta-khana* in particular. Contemporary scholarship holds that 'during Akbar's time cheetahs were classed into eight categories and food was rationed according to class. The first received 5 seers (one seer = 933.12 grams) a day of meat… Eighth, 2¾ seers.' Based on this calculation, on average, 1,000 cheetahs required (at a minimum) a daily supply of 3,610 kg (kilograms) of meat or 1,317,650 kg of meat annually. The *Ain-i-Akbari* further divides cheetahs into *misl*(s), where each *misl* comprises ten cheetahs. Only three *misl* of *khasa* (special) and two *misl* of a lesser category were kept at court. Thus, while 50 cheetahs attended the emperor at any given time, the other 950 were apparently kept 'in His Majesty's park.' Each *khasa* cheetah was allocated up to six keepers, thus the 30 cheetahs in this category at court required the presence of 180 attendants to 'service' them. The job of the *darogha* (store superintendant) required him to enter the food consumption of each cheetah in his *Rozna'amcha* (Day Book) and this was randomly checked by the emperor himself. The feeding followed a strict protocol under the *pa'gosht* regulations, which was a determined quantity of feed relative to the weight, age and size of each animal. Once in the field, the cheetahs came under the supervision of the Mir Shikar (Lord of the Hunt) of that particular hunting ground.

The royal Mughal emblem, represented by the lion and the sun. The lion was an integral part of their being, as it was for rulers across the world.

Furthermore, if we accept the premise that Akbar captured

even ten per cent (a huge number) of the total cheetah population that was apparently roaming the plains of India, 1,000 cheetahs in captivity would mean a further 9,000 felids in the wild. The big cat scientist, Raghu Chundawat, calculates the annual consumption (in terms of predation) that such a population would require:

> [Basing my calculations] on the per day biomass consumed by captive cheetahs, a cheetah in the wild would consume between 2.5 to 3.5 kg per day in most cases (minimum around 0.8 kg). If we take this information to calculate the yearly consumption...[the] biomass requirement for one cheetah is 1,095 to 1,277 kg. Captive animals are fed dressed meat, whereas in the wild the situation is different. Therefore, I added 30% in waste as bone and skin. This amounts to 1423-1660 kg of biomass required for one cheetah. I converted this into blackbuck units (BB units) and each BB unit is estimated on average at 35 kg (adult BB weight ranges from 30 kg to 40 kg for males; since the cheetah predates on various age and sex categories such as females, sub-adult, juveniles and adult males, I calculated a BB unit to be at anywhere between 25-40 kg). Based on this, I estimate that the cheetah would have had to predate on 47-55 BB units in a year. When we assume that the cheetah is harvesting at a rate of 10%, then approximately 470-550 BB units are required to sustain this predation. Hereon, it is a simple multiplication: for 100 cheetahs, 47-55,000 BB units; for 1,000 cheetahs, the requirement would be 470 to 550 thousand. For 10,000 cheetahs, it would be 4.7 to 5.5 million BB units.

Therefore, to sustain themselves, 9,000 cheetahs in the wild would require a stable population of approximately 4.23 to 4.95 million blackbucks within the boundaries of the Mughal Empire. Other prey species which may have supplemented an assumed non-availability of the blackbuck, such as gazelle or hare, would have had to exist in equally massive numbers. Moreover, the cheetah was certainly not the only predator on India's grasslands and plains. Hyena, wolves, jackals, wild dogs and the occasional tiger and leopard would have competed with the cheetah for every available morsel of food. A human population of anywhere between 120-130 million (as it was in Akbar's time) would have exacerbated the problem. Further, the scientists' calculations do not take into account the royal hunts carried out by the Mughal emperors and other notables. Recalling one of his more restrained hunts, Akbar's son and successor, the emperor Jahangir, records:

> On the first day, I went to the village of Samonagar, which is one of my fixed hunting places to hunt. Twenty-two antelope were killed, of which I myself killed sixteen and Khurram the other six... I ordered that *ja'i-namaz* (prayer carpets) should be made of the skins of the antelopes I had myself killed, and be kept in the public audience hall for people to use in saying their prayers.

Only a few days later, as if to compensate for the meagre game bag he had secured, Jahangir takes to the field again, only this time he is also preoccupied with sustaining the deer population in the imperial hunting grounds at Fatehpur Sikri and resorts to the novel method of relocating them there from other reserves. He observes:

> When news came that the hunting place had been prepared and a great deal of game had been confined, I went there and began to hunt on the Friday... Some of the deer were taken alive and some killed with arrows and guns. On the Sunday and Thursday, on which I do not fire guns at animals, they took them alive in nets. In these seven days, 917 head, male and female, were caught, and of these 641 deer were caught alive. 404 head were sent to Fatehpur to be let loose on the plain there, and with regard to 84 I ordered them to put silver rings in their noses and set them free in the same place. The 276 other antelope that had been killed with guns and arrows and by cheetahs were divided from day to day among the Begams and the slaves of the palace, and Amirs and servants of the palace. As I became

> very tired of hunting, I gave orders to the Amirs to go to the *shikargah* and hunt all that were left over, and myself returned in safety to the city.

This event dates to 1611, barely six years after the death of Akbar. No tally is given of the number of antelope killed by the nobles or of how many of the relocated antelope actually survived. If there were cheetah in the wild, other predators were evidently not the only entities straining the prey base. Between the ages of twelve and fifty, Jahangir had accounted for 1,672 blackbucks, at the rate of 44 every year. Barely three years after he had calculated this game bag, the emperor recorded catching 426 antelopes in a space of just twelve days at the Palam *shikargah*. The cheetah was fast becoming one of the catalysts in the destruction of its own potential prey base.

▪

This brings us, inevitably, to the status of wild cheetahs at the time of Akbar and, subsequently, his descendants. Given what we know of existing hunting reserves—stocked with animals both imported and 'indigenous'—in ancient India and the absence of any evidence of cheetah remains, whether in fossils, cave paintings or seals excavated around ancient settlements, is it a plausible hypothesis to suggest that the cheetahs Akbar encountered 'in the wild' had either been let loose for breeding or were escapees from the menageries and hunting grounds of previous monarchs, which in some cases had turned feral or semi-wild? Could they have been part of the 950 cheetahs, separated from the 50 kept at court, some of which were, in any case, considered wild? The few instances in the *Akbarnama* which describe Akbar's personal capture of cheetahs occurred in or near imperial hunting grounds, suggesting that the 'captured' cheetahs were either escapees or feral animals. In 1560, the first one was caught in Hisar-Firuza (in present-day Haryana), a hunting reserve from the time of Firoz Shah Tughlaq, after whom the town was named. The exact location of the second hunt is not given, though it was captured near Gwalior in 1569, which had imperial hunting grounds all around it, most notably the Bari-Dholpur grounds and one at Khanua, both known for their game. In another incident, the ever-faithful Abu'l-Fazl attests that:

> …no less than seven cheetahs were caught. At the time of their heat, which takes place in winter, a female cheetah had been walking about on the field, and six male cheetahs were after her. Accidentally she fell into a pit, and her male companions, unwilling to let her off, dropped in one after the other—a nice scene indeed.

And possibly a fictional one. The location of the incident is not clear, though we can hazard a guess that this too may have occurred in or near an imperial hunting reserve stocked with cheetahs. How else could six male cheetahs, whose natural behaviour does not suggest sociability (except between mother and cubs and that too for a limited period), be tolerating each other, especially around a female? The story is more amusing than insightful. Cheetahs, both male and female, range over vast territories and the location of a pit trap exactly on the path of seven carousing cheetahs seems to be too much of a coincidence. Like the other accounts in Akbar's time, it does little to provide proof of an Indian provenance for the cheetah.

During a Pattan hunt in 1571, six cheetahs were caught in or around the Dipalpur hunting grounds. Why is it that we don't hear of cheetahs captured away from the hunting reserves and grounds especially created for royal sport? A clear pattern is apparent in the controlled, almost enclosed location which limited the spread of the cheetah population beyond areas earmarked for their breeding. For, as we know from Mughal accounts and others, cheetahs seldom breed in captivity.

Despite the elaborate protocols, the delegation of massive resources (both human and material) and the value the emperor himself attached to the upkeep of his cheetahs, even Akbar could not prevent cheetahs escaping their royal bounds. A particular incident is described in the *Ma'athir-ul-Umara*, while relating the biography of Raja Anup Singh, the famous Sinh-Dalan of Jahangir's reign, and is revealing of a larger theme:

> ...his grandfather on account of poverty used to hunt deer, and live upon their flesh. By chance he one day in the jungle fired at what he thought was a tiger. He hit a royal *cuta* which they had let loose at the deer, and which had secretly entered the jungle. The bell and golden collar enabled Anup Singh's grandfather to recognize that it belonged to the royal establishment. He took off the trappings and flung the body into a well. Those who were looking for the *cuta* came to the well and gathered that this was the work of the Rajput who was always going about hunting. They went to his house and got the bell and collar. They also seized him and brought him before Akbar. When he was told what had happened, he approved of his courage and marksmanship and took him into his service. On account of his love for shooting, he gave him a suitable office.

The story is interesting even as fiction because it would have been based upon plausible occurrences its chronicler would have witnessed about him in real life. As animals trained for the hunt, cheetahs were able to survive in relic, feral populations, but even this could be hazardous. Nor was this the only way in which cheetah casualties piled up. Mishaps also occurred while the royal pets were being transported. For instance, two of Akbar's favourite cheetahs, Daulat Khan and Dulrang, were drowned while crossing the Ganga during one of his easterly expeditions.

Annemarie Schimmel, in her book, *The Empire of the Great Mughals*, makes an intriguing assertion that cheetahs had become thoroughly 'acclimatized', especially in Gujarat. The allusion to a procedure of acclimatization in Gujarat and, therefore, close to the imperial port of Surat, suggests that cheetahs (among other 'commodities') were possibly shipped in from Africa. Mughal annals frequently feature the exchange, imports and exports of animals, exotic as well as mundane.

With the passing of Akbar, the 'golden age' of the cheetah in the history of India, briefly splendid and more enchanted than the life of any other comparable species, began an irreversible decline.

Cheetahs would continue to be used for the chase and foreign visitors would still peer at them in the various menageries of the 'kingdoms' that succeeded the Mughals, escapees would continue to turn feral and be killed for 'sport' and so on, but the animal would never come close to enjoying the status bestowed on its ancestors such as the legendary Citr Najan at Akbar's court which is said to have cleared a ravine of 25 yards in a single jump when it was chasing a deer; it finally caught its prey, to the delight of the emperor, who '...made him chief of the *cutas*. He also ordered that as a special honour, and as a pleasure to men, a drum should be beaten in front of that *cuta*.'

The *Ain-i-Akbari* further recorded that the *khasa* cheetahs received brocaded saddle cloths, chains studded with jewels and *ghushqani* carpets to sit upon. It talks about the royal treatment meted out to another favourite cheetah, Samand Manik, who was carried on a *chau-dol* (palanquin) and accompanied by much pomp and show. 'His servants, fully equipped, run at his side; the *naqqara* (war-drum) is beaten in front, and sometimes he is carried by two men on horseback, the two ends of the pole of the chau-dol resting on the necks of the horses.'

▪

Jahangir, the fourth Mughal emperor, was an enthusiastic observer of nature. Among all the rulers of India, he is perhaps the one who is most qualified to merit the distinction of an amateur naturalist. Usually inebriated by alcohol and drugs, the temperamental king always found the time to record his findings about mammals, reptiles, birds, insects and the flora he encountered. A relentless hunter and experimentalist who would study animal parts and entrails to ascertain their habits and physical characteristics, Jahangir's discoveries regularly found their way into his memoirs.

> Foreign dignitaries, his own nobles, hermits, traders and commoners brought for him presents of animals from far and wide... His intimacy with nature was far more extensive than the Tuzuk reveals.

Jahangir also had a network of agents whose principal occupation was to source exotic species. He would send these men to various corners of his empire and even abroad to procure species for his collection. In one instance, Muqarrab Khan, a Mughal grandee and confidante of Jahangir, was commanded, '...to go to the port of Goa and buy for the private use of the government certain rarities procurable there... When he returned from the aforesaid port to the Court, he produced before me one by one the things and rarities he had brought. Among these were some animals that were very strange and wonderful, such as I had never seen, and up to this time no one had known their names.'

At the conclusion of another assignment, the same Muqarrab Khan returned with a pachyderm from Africa, a venture which excited less glee because the import of African elephants had occurred before. Yet it was important enough to merit a mention in Jahangir's private journal:

> Muqarrab Khan presented an offering of a small elephant from Abyssinia which they had brought by sea in a ship... In the time of my revered father, Itimad Khan of Gujarat sent a young elephant as an offering...

At another time Jahangir notes, 'On the 12th the offering of Khan Dauran, which consisted of forty-five horses, two strings of camels, Arabian dogs, and hunting animals, was brought before me.'

Even late in his life, Jahangir was receiving animals to add to his collection. In 1621, his journal records: '...the offering of Lachmi Chand, Raja of Kumaon, consisting of hawks and falcons and other hunting animals, was brought before me.'

No presentation of animals of the chase caused more excitement than one particular gift from Jahangir's chief henchman, the brutal and conniving assassin of Abu'l-Fazl himself, Raja Bir Singh Deo Bundela. This forms the only recorded instance of a white cheetah in history and we learn nothing of where or from whom Bir Singh Deo Bundela acquired it. Writes Jahangir:

> Raja Bir Singh Deo brought a white cheetah to show me.

> Although other sorts of creatures, both birds and beasts have white varieties... I had never seen a white cheetah. Its spots, which are (usually) black, were of a blue colour, and the whiteness of the body also inclined to blue-ishness.

All the Mughal emperors received gifts of 'animals of the chase' and other beasts presented to them by vassals, defeated rebels and foreign dignitaries. What were these 'animals of the chase', these *janwar-i-shikar*, which keep turning up without scholars making more than a cursory attempt at finding them in the records of the Mughals? The various traders, agents, embassies and vassals who presented them were unlikely to be presenting ordinary beasts to the fabulous Mughals. Anthony Cutler's argument sheds some light on this process of gift-giving by expanding on Alfred Marshall's theory that the value of an object increases in proportion to its scarcity:

> Given that the sources understandably concentrate on offerings by and to emperors, caliphs, and kings, stressing their rarity and in many case their uniqueness, it is hardly surprising that gifts have come up against such notions as marginal utility.

▪

The consequence of this massive movement of 'animals of the chase' between cultures and empires was that hunting animals took on a cult status with international elites across the world. Akbar's obsessive reliance on the cheetah as a hunting companion gave way to Jahangir's preference for hunting dogs. While he continued to import them, and even hunted with them occasionally, Jahangir was also disbursing the cheetahs painstakingly collected by his father. On one occasion, he records presenting five cheetahs to Qasim Khan, the Mughal governor of Bengal, a significant gift considering how much Mughals valued the cheetah. There is no record of how this set of cheetahs fared, if ever they found their way to Bengal. Nor was this dispersal confined to the boundaries of his empire. Munshi catalogues

Jahangir's embassy to Persia in 1620 comprising of:

> ...ten huge elephants equipped with gold howdahs and embellished with all kinds of trappings, and a variety of animals, including tigers, leopards, antelope, Indian lambs, cheetahs, rhinoceroses, talking birds and water-buffalo which pulled various types of litters.

In fact, Jahangir once even managed to kill a cheetah and a caracal, a trophy actively sought by princes because these animals were aids for hunting, not prey. Ever vigilant about recording his game bag, the emperor recorded in his own hand that during his visit to Ajmer in 1616, 'I went out to hunt tigers, etc., fifty times. I killed 15 tigers, 1 cheetah, 1 black-ear [caracal], 53 nilgaw, 33 gazelle, 90 antelope, 80 boars, and 340 waterfowl.'

The ratio of tigers to cheetahs killed is a telling one and it is improbable that an accomplished hunter like Jahangir would have accidentally killed a cheetah.

▪

While Akbar was responsible for elevating the status of the cheetah, the lion was already embedded in the Mughal consciousness as a royal symbol from earlier times. This was again, a consequence of Timur's adoption of the ancient symbols of Iran after his conquest of Samarkhand and the conversion of that great city into his own capital. A contemporary description of Timur's palace, especially of the ornamented doorway that led to court, describes the origins of this development:

> On the top of this doorway there was the figure of a lion and a sun, which are the arms of the lord of Samarcand [sic]...because the sun and lion, which are here represented, are the arms of the lords of Samarcand [sic]...

In the centuries to come, Timur's descendants would carry this symbol on their flags to as far as Bhutan in the east and Masulipatnam (now Machilipatnam) in southern India. Further support for the Iranian origins of the Mughal lion, both symbolic and in the flesh, is found in the following extract:

> The Persian lion is now extinct in Iran, and there are no confirmed modern records of lion presence in central or eastern Iran, or Baluchistan, but it's believed that lions that still live in India are the same lions that once were living in Iran. The symbol of the old flag of Iran is a lion holding a sword in his hand and with a half of the sun behind him… The Persian lion once lived in the valley of Dasht-e-Arzhan as well as the 'Kam-Firuze' and 'Gourab' hunting ground, south of Hamedan. It used to roam the oak forests of the Zagros Mountains and riverine areas of Khuzistan.

Wherever Mughal armies marched, they carried upon their flag the symbols of a lion couchant, with the sun rising behind it. This was another classic appropriation of a rare and magnificent species that was largely unavailable to public viewership but widely used as an esoteric symbol of royal prestige. As we have seen from contemporary historical accounts, the Mughal lion was seldom observed in India beyond the vicinity of royal hunting reserves. Confusion about its provenance and presence in the Mughal period has arisen from the interchangeable use of the term *sher* or *shir* in the Mughal narrative, which at times denoted a tiger and at other times, a lion. Sometimes (though rarely), these terms were used—and continue to be used colloquially—as a generic description of any large cat, even the leopard. The *Tutinama*, commissioned during the reign of Akbar, features both tigers and lions in its stories, but uses the generic term *sher* for both.

While Mughal artists were generally accurate about depicting their subjects, the captions and accompanying literature perhaps betray a lack of understanding about the subject they were painting. However, sometimes the imagination could impose upon the reality of the depiction. During the Mughal reign, artists were always careful about denoting a scene as truthfully as possible but as art historian Som Verma states, '…in such narratives, the descriptive details in the visual field are expanded by the artist by drawing upon his experience though imagination, too, played a part.'

Colloquial and common usage of names for four of the big cats during Mughal times (as well during the Safavid and

The Mughal painters could depict either lions or tigers in their representation of a specific event, causing much confusion in historical records. Here, Anup Rai fights off a lion.

Ottoman empires) did little to make things clearer. The following table shows just how confusing this could sometimes be, particularly as Urdu terms came into circulation towards the time of the later Mughals:

	Arabic	Persian	Urdu
Lion	*asad*	*sher*	*babar-sher/babar*
Tiger	*namr*	*babar/sher*	*sher*
Leopard	*namr*	*palang*	*cheeta*
Cheetah	*fahd*	*yuz palang*	*cheeta*

A couple of contemporary accounts from the period illustrate this confusion well.

The *Tabaqat-i-Akbari* narrates:

> At the beginning of this year the Emperor left Ajmer, and proceeded by the way of Mewat towards Agra. On his

The same event of Anup Rai hunting depicted here too, but with a tiger.

journey, he passed a jungle which was the abode of lions (*sher*) and tigers (*babar*). A terrible lion (*sher*) came out, and His Majesty's followers, who were constantly in attendance upon him, discharged their arrows and stretched him in the dust. His Majesty then gave orders, that if a like thing should occur again, they were not to shoot until he directed them. As they went on, another lion (*sher*), larger and fiercer than the first, came out and made towards the Emperor. No one of the attendants dared to fire without orders. The lion-hunting King alighted from his horse and levelled a musket at the beast. The ball grazed the animal's face, inflicting a slight wound, which caused him to rush from his place towards His Majesty. The Emperor fired a second time, and brought him down. At this juncture, Adil Muhammad Kandahari boldly placed an arrow to his bow, and faced the animal, which then turned away from the Emperor and attacked him. It brought him to the ground, and was about to take his head in his mouth. That brave fellow, in this supreme moment, thrust his hand into the animal's mouth,

> and sought to draw his dagger to stab him in the belly. But the handle of the dagger stuck in the sheath, and the beast gnawed the flesh and skin of the hand which was in his mouth. Notwithstanding this, Adil managed to draw his dagger, and inflicted some deep wounds in the animal's belly. Brave men gathered round on all sides and finished him... After the lion hunt the royal camp moved towards Alwar...

The same story is given in the *Akbarnama* but the translations do not make a distinction between lion and tiger as this one does. However, the translator, Sir H. M. Elliot writes in his notes: 'It was in all probability a tiger, although the author would seem to use the words *sher* and *babar* distinctively.'

Similar usage continued into the reign of Jahangir. Marvelling on the stripes of an imported zebra, the emperor commented:

> At this time I saw a wild ass (*gur-khar*), exceedingly strange in appearance, exactly like a lion. From the tip of the nose to the end of the tail, and from the point of the ear to the top of the hoof, black markings, large or small, suitable to their position, were seen on it.

The editor's notes on this observation are interesting. Jahangir's 'wild ass' is 'apparently a zebra. See *Iqbalnama*, 179, where it is stated that it was brought by sea. The text of the *Tuzuk* is wrong, as usual. What we should read is: "It was like a tiger (...*Iqbalnama* [has] *shir*, not *babar*), but the markings on a tiger are black and yellow, and these were black and white."'

The editor is being needlessly pedantic about this particular point though—the zebra's black stripes are evidently being compared to the stripe patterns on a tiger's body, which appear similar. He is correct to assume that a tiger (not lion) is meant, even though the text has *sher*. All this evidence must necessarily question the accounts which make no distinction between a lion and a tiger. Consequently, this has led to speculation and assumptions about the lion's provenance based on Mughal records.

Misappropriation of nomenclature apart, the Mughals *did* hunt lions, and did so grandly and with precisely managed

The zebra was talked of as a lion but was meant to be a tiger because of its stripes. Emperor Jahangir received this one from Abyssinia (1621).

arrangements. As it was for monarchs before them, the Mughals considered the killing of a lion auspicious, its escape in a hunt a very ill omen indeed. There was something near-mystical about the lion hunt, a practice entangled in superstition, myth and fear. Perhaps the Mughals were directly influenced by the Persians who warned:

> You must know that there is no sport more difficult than that of the saker and crane, but there is also none better; none, except that of the lion with a buffalo, and the cheetah with a gazelle. I have hunted many a lion and seen trouble therefrom, for the sport is inauspicious; for the lion is the King of Beasts, and His Highness the Commander of the Faithful (on whom be peace) is styled the Lion of God. Hence the sport of the lion is baleful, and he that follows it will certainly see no good; still it's a fine sport; I have tried it—but my advice to you is on no account to do so, else you will regret it, for no benefit accrues therefrom.

The total number of lions killed were few, and often far between. The combined reigns of all twelve Mughal emperors hardly add up to more than a few dozens, keeping in mind that often enough it was the tiger chroniclers spoke of, rather than the lion. As we know, the prize of killing a lion was reserved for the emperor, who could, if he wished, bestow the honour on a particular individual. As a commodity, the lion was rare, highly valued and generally protected in areas reserved for hunts.

The hunt itself was given enormous importance. We will see more of this in the next chapter but just one instance of the high regard the Mughals had for it—whether as an apex of leisure or as a preparation for war—is illustrated by their high esteem for the office of the Mir Shikar, or Master of the Hunt. As a veritable Department of State, the Shikarkhana or Department of the Hunt, had its own equipage, an army of scouts and huntsmen who not only looked after the health of the imperial hunting grounds or *shikargah*s but also tracked and marked down quarry for their royal masters. This department also maintained the vast royal menageries, following strictly enforced guidelines and procedures for the grooming, feeding and transport of royal animals. Like the animals in their charge, the huntsmen were frequently imported or exchanged between countries and often formed part of imperial delegations travelling overseas. The memoirs of Jahangir refer to numerous instances of huntsmen being honoured. This was only natural, for certain races specialized in the training of (and coursing with) particular species that originated from the same homeland as the huntsmen themselves. Nonetheless, the imperial huntsmen were an important presence at the Mughal courts. For the emperor's private pleasure, they were a critical necessity. The nominal charge of a grandee to oversee these clusters of huntsmen was usually a ceremonial position that honoured nobles at court. On top of this gigantic imperial superstructure sat the emperor himself, wielding his weapons of the hunt.

The Mughal hunting grounds and parks were scattered all over the empire. They could be found along old military routes and sea ports and whichever areas the emperors believed

A giraffe being given as a tribute at the Mughal court in the seventeenth century. How many African animals were imported by kings and emperors into India?

resembled the natural habitat of the lion and the cheetah. Parks were located near Pattan, Bhatnair, Bhatinda (in the present day states of Punjab and Haryana) and throughout Rajputana and Marwar around Amarsar, Dholpur, Jodhpur, Nagaur, and Mertia—core territories of the empire.

Around the time the Mughal Empire began disintegrating and the British Raj had not quite gained complete ascendancy in parts of India, the Mughal trend of regenerating areas into forests for the purpose of the hunt was being continued by the successor states of the empire. Observers noted how Sindh proved to serve as a good example of how the East India Company took over state forests and hunting reserve systems and ran them under their own laws:

> Between 1690 and 1830 the Amirs of Sind were responsible for the reafforestation of over a million acres of the Indus flood plain with up to eighty-seven *shikargah*, or hunting and forest reserves, whose forest products were sold to the peasantry.

As a symbol, the uses of a lion varied. Aurangzeb, for example, enjoyed the metaphor of the lion and a goat and 'as proof of his justness and to advertise his good deeds he would set out every day to walk through the principal square a fierce lion in the company of a goat that has been brought up alongside it from birth. This is to show his decisions are just and equal without any bias. This is done in court solely because the world may be notified of his justice.' Such were the uses of the 'tame lion' bred from birth with goats!

Manucci observes that the emperors were very competitive and went to great lengths to convey their lion stories to each other, not always to cultivate friendly relations. As we are told by reports, when Aurangzeb's ambassador, Tarbiyat Khan, was received at the Persian court of Shah Abbas, the Safavid ruler of Persia, '...Shah Abbas repaid the hauteur of Aurangzeb in treating his ambassador badly. Among other pleasantries, one day when his ambassador was at the audience, the king ordered into his presence a lion secured by two chains of silver gilt. When the lion appeared, he took hold of it by the mane and stroked it, to show how brave he was. The lion, which was tame, let itself down gently on the floor, and made friendly gestures to the king. The ambassador was in a wonder, and Shah Abbas, as a joke against the reporters (accompanying the ambassador), said: "Write this too, to Aurangzeb."'

In the following sections of this book, Valmik Thapar retraces our journey of the Mughal annals through the eyes of later visitors to the empire and India. He recounts in detail the elaborate preparation and ceremony of an imperial Mughal hunt, as witnessed by starry-eyed, sometimes even envious foreign observers. The conclusions of these chapters are elaborated upon using modern accounts, his vast experience studying mammals and (particularly) big cats in the field over the last four decades and a vast narrative archive of the British Raj in India. To conclude our own journey alongside the cheetah and the lion through the rich landscape of the Indo-Islamic empires, it would suffice to say that these dynasties gave the two species a prominence and permanence in India's history that they had not enjoyed before. Delhi's sultans and the Mughals after them, inadvertently (or indeed by design), embedded these exotic aliens into India, sometimes as companions, at others as competition worthy of their prowess so effectively that the cheetah and the lion were no longer alien, though they would always remain exotic.

This eighteenth century painting shows a lion attacking an elephant as it is beaten off by the hunter. Again, I believe this was how the Mughals wanted to publicly display their power over the lion and the painters reflected events that may or may not have occurred. And such encounters when they took place must have been in huge private hunting parks that were carefully managed.

Next page:
A man grapples with a lion as must have happened for thousands of years, to represent the quest for power over the most powerful of animals.

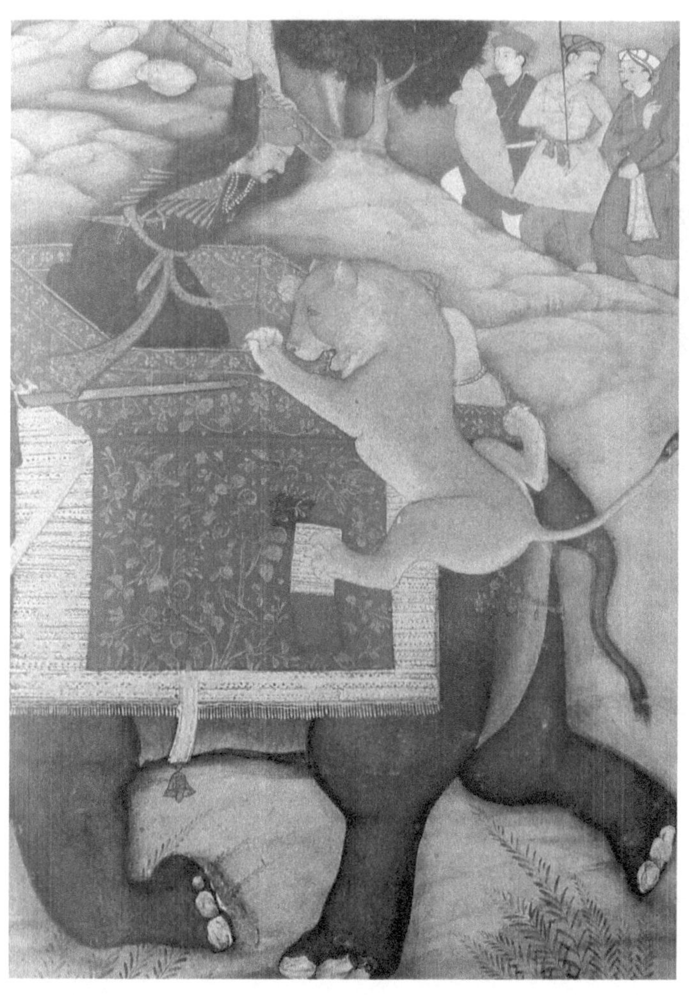

IN THE COURTS OF EMPERORS

by VALMIK THAPAR

When I first saw so-called Asiatic lions in the mid-1990s in Gir, their behaviour resembled that of large dogs. As I walked along a path with a forest guard, watching them from a few feet away, other guards hissed, hooted and mimicked endless varieties of domestic livestock to attract their attention. The spectacle did not even remotely resemble what I expected to see when it came to encountering wild lions. Indeed, I do not think I have ever seen a more disappointing sight in the wild. The tameness of these animals made me recall images of lions stuffed with opium (and wild asses) until they were lethargic enough to be slaughtered by Mughal emperors for sport. I then had to wait several years to see the lion in Kenya. That experience was exactly the opposite—I wouldn't have dared to get out of the relative safety of my vehicle, let alone walk around them. In the Masai Mara, I saw more than a hundred different lions in ten days. These were 'real' lions and not the caricatures I had encountered in Gir.

The suspicion that had been sown in my mind during my research for *Tiger Fire* (to be published in 2013) now emerged as full-blown theory. It was clear to me by now, through the books I had consulted as well as my encounters on the ground, that the Indian lion was never a native species or subspecies but had always been an exotic import that had been shipped into the country and bred for sport. As there is no conclusive genetic evidence to prove or disprove this thesis, I've had to necessarily rely on the historical record and my own training as a naturalist.

Let's start with habitat. This country's dry, moist and evergreen forests make for a very unwelcome terrain for lions, especially given the presence of tigers, the more powerful predator. Even where there are grasslands or bush (the lion's natural habitat), I couldn't imagine them eating or feasting as a pride on cinkara and blackbuck. Maybe nilgai, the largest Indian antelope, could have been suitable prey for the lion, but if that was the case, why was there were hardly any records of encounters between the nilgai and the lion? The fact remains that if the nilgai was not the lion's natural prey, the country simply did not have the millions of blackbucks that would have been needed for 'indigenous' lions and cheetahs to feed on. In Africa, however, there are more than 40 species of antelope of different sizes as well as zebras, giraffes, wildebeest, wild buffaloes and warthogs—plenty of prey for the lion, which is why it flourishes in that continent. It was also odd that there were so few records of encounters with prides of lion in this country; in Africa, a pride can consist of as many as 30 animals.

What is apparent on the ground is true of the historical record as well. Romila Thapar and Yusuf Ansari have dealt with the lions in antiquity and in the Islamic and Mughal eras, but some of the points they make bear reiterating. The Mohenjodaro seals do not depict the lion, nor do early cave paintings that date back 5,000-8,000 years ago, whereas the tiger is to be encountered everywhere. Some observers argue that the lion was a late entrant into the subcontinent, and that it only arrived on the scene 2,000-3,000 years BCE. I find that difficult to imagine because it is inconceivable that lions could have swum across the Indus (unlike tigers, they hate water) to make a home for themselves in the plains of Delhi and Haryana. Moreover, experts have testified that there '[existed] absolutely no proof of the occurrence of lions' in Afghanistan and Baluchistan—regions that bordered India. If this was indeed the case, how could wild lions have migrated into the subcontinent without passing through lands that lay just across the border? The only explanation that fits—and it is one we will examine in detail in this and subsequent chapters—is that the lion was imported into the country in large numbers and bred for release into the designated hunting grounds of Indian royals.

We have evidence of traffic in live lions from approximately the third century BCE up until the fifth century CE (especially towards Rome from different regions of Africa, much as tigers were sent from India) but for all we know, the trade in lions went back much further than that because the animal has been associated with royalty from time immemorial, and had, therefore, always been in demand. As Thomas Allsen writes:

> Ali Akbar Khitai, a Persian merchant who wrote an invaluable book on China in 1516-17, says that Muslim traders travelling overland often brought lions (*shir*) and other cats to China where they fetched a good price... The lion in China far from the natural range, had from the beginning great symbolic power.

Another clue as to the origin and place of the lion has to do with the fact that from Alexander's times (third century BCE), accounts of the animal have often talked about how tame it was. Should we take this to mean that throughout its history in the subcontinent it had always been captive, tamed, fed, trained and then sometimes hunted? We know that in the twentieth century when the lion had almost become extinct, the Nawab of Junagadh decided to manage the population in Gir (his then designated hunting ground) through intensive artificial feeding and protection. Did the Mughal emperors do the same in their time? Was the 'Royal Lion House' a breeding and holding station to train lions for the court and release them into the stage-managed hunting grounds of the emperors? If Jahangir boasted 100 lions in his court, how many did he have in his stables?

▪

We know many ancient cultures set great store by the lion because it was emblematic of royalty; in this, they were merely following in the path of a well-known historical tradition.

> It was because men dreaded the lion that he became the emblem of wisdom in Assyrian sculpture and the type of courage in Hebrew poetry; that his head crowns the body of an Egyptian god, and that his form has been taken as a royal cognizance in the East and West. For no other cause is it that death is the penalty for anyone but a ruler to wear his claws in Zululand, or that among the Algerian Arabs his whole body possesses magic virtues.

Indeed, it is hard to find a culture in which the lion was not revered by the ruling class. It was deified by the people of the East—a part of the pantheon of gods in Mesopotamia, it was also considered sacred in Phoenicia. Leontopolis and Heliopolis in Egypt were central to the cult of the lion; in Amun Ra's temple at Heliopolis, 'lions were tended carefully by high ranking priests who bathed them in perfumed water, salved them, burnt incense for them and fed...them the choicest tidbits to the strain of sacred music.' So much so that laws were made decreeing a public mourning on the death of one of these animals, which would then be 'embalmed and entombed with great ceremony.' The lion was closely associated with the goddess Cybele—whom the Romans knew as Magna Mater—and tamed lions were kept in temples built for her. 'In ancient Greece the priests of Cybele who wandered across the country like mendicant monks, frequently had a lion with them, and made use of it in their exorcising rituals.'

Just like Mughal emperors, the pharaohs liked to appear in public in the company of tame lions; even when Ramesses II went to battle, he was accompanied by a lion that ran alongside his chariot and knocked down anyone who tried to approach the ruler. In peacetime, too, lions were constant companions to emperors and kings. Tame lions were kept at Leontopolis, the lion city in the Nile delta and near modern-day Cairo, where they were propitiated.

> Because of their long association with royalty, with royal crests, valour and so forth, there was never any challenge to the lion's status as King of the Beasts. In the hierarchy that was the animal kingdom (as then perceived), the lion's right

to the title went without question.

In Baghdad, the Abbasid Caliphate would impress guests with 100 lions suitably collared and muzzled and with a keeper for each. This was in 917 BCE.

Observers believe that Ashoka's empire in India was deeply influenced by Persia and Persian-trained Greek sculptors who were involved in the creation of the extensive lion art that exists in India. According to M. A. Rashid and Reuben David: 'Persian influence in architecture during this time shows Achaemenid influences through the elongated petals and realism in the deeply carved snarling lions and also other animals which are similar to the Persian Persepolis designs, which proves that the lions migrated into India via Persia.'

Some scholars (as do I) doubt the veracity of this argument to account for the presence of lions in the subcontinent. What is apparent, though, is that lion art flourished in areas where lions were never found. Probably the most well-known connection to lions in India is the depiction of the lions on Ashoka's pillars. Romila Thapar and other serious historians argue that the most familiar visual connection with the Ashokan pillar and its lions is found in the region of Persepolis around Susa and Iran. Other scholars like Saryu Doshi believe 'that Egypt inspired the Ashoka's conception of the Sarnath pillar bearing the famous lion capital.' Whatever might be the truth about Ashoka's lion pillar, one fact is certain—the lion in it did not come from

An Etruscan stone relief from the sixth century BCE depicting lions on a leash—tamed to be presented in court or being readied to be released for the hunt.

lions in the natural landscape of India but from connections across the seas. The other fascinating fact of this early art is that there were very few depictions of lions attacking prey or humans; they were always shown in guardian-like postures that appear to be more symbolic than anything else, unlike the depictions of the animal in Assyrian art or even at Persepolis.

Left: *The lions on early Pharaonic pillars were remarkably similar to the lions on the Ashokan pillar.* Right: *The Ashokan pillar reportedly got its major influences from outside India.*

It doesn't help matters that, because of royal privilege, anything to do with the lion was always kept a secret and is one of the reasons why there are so many missing links in our understanding of this species.

Every emperor from China to India to Europe and across Persia and more or less everywhere in the known world was in the market for lions. In 1592, a lion and lioness in the Tower menageries in England were even named after King Edward and Queen Elizabeth I. King James in 1622 owned 11 lions. Hunting parks blossomed across Western Europe and even if European royalty didn't manage to shoot that many lions and tigers, they managed to keep themselves busy setting up parks for their animals.

▪

The hunting of lions in antiquity was deemed the sport of kings. In 669-630 BCE, the Assyrian king, Ashurbanipal, was known for using rather gruesome techniques to hunt lions. As the king and his retainers waited on horseback with arrows and spears, lions would be released from their cages; while spectators watched the scene unfold from a hill nearby, the king would go after the

In private hunting parks, lions were released from wooden cages some 2,800 years ago, a tradition that continued to be used in lion hunts for years to come.

enormous male lion. The animal stood no chance:

> Paralysed by arrows that have severed her spinal cord, a lioness drags her hindquarters along the ground, determined to face her attackers despite her injuries. Nearby another dying lion vomits blood. Spearmen and dog-handlers with mastiffs guard the perimeter of the hunting area to ensure that none of the animals escape.

Pictorial depictions of Mughal hunts resemble Assyrian hunts that have been carved in stone. Guggisberg confirms this:

> The Moghul Emperors of 17th and 18th century India staged lion hunts of a splendor and scale rivaling those held at the time of the Pharaohs and Assyrian Kings.

Aurangzeb would surround a whole region with his armies, and over several weeks (if not months), stage the most 'fantastic massacres'. Guggisberg thinks 'Indian princes may occasionally have trained lions for the chase, as Pharaohs had done long before them.'

▪

Before the Mughals arrived on the scene and began showing off their lions, there are hardly any descriptions of lions in the accounts of early travellers to India. Sung Yan, early in the sixth century CE, only sees lions in what was probably Turkestan and

This kind of canned hunt in enclosed areas (seventh century BCE) continued into the Mughal era and well into the middle of the twentieth century.

that too as tributes being sent to the King of Gandhara. He made this observation after he had traversed the whole of middle India. Another record in the same century reiterates this movement: 'A small state in Turkestan possibly Balkh, sent lion cubs to Gandhara in northern India in the early sixth century.' Bactria or Balkh in Turkestan seems to have been the centre from which lions were sent either to India or China. Even till the thirteenth century, Marco Polo says, 'There was plenty of game around Balkh and there are lions too.' This is where Alexander is said to have killed a great lion north of the Oxus towards Samarkand. Timur also killed two lions in the same area more than 1,000 years after Alexander.

Ibn Batuta, who travelled through India and Sindh from 1324 to 1354, never talks of lions but mentions the rhinocerous. Early travellers to the Mughal Empire, such as Sidi Ali Reis, are silent on the subject of lions in the wild, which is corroborated by several others. Father Monserrate never sees wild lions on his journey to the court of Akbar between 1580 and 1582.

The Dutchman, Linschoten, is very clear about the absence of lions in sixteenth century India. He writes:

> Within the lande there are also tigers; other cruel beastes, as lions, beares and such like there are [fewer or] none.

Linschoten is also one of the first travellers in this period to

Top: *This is how lions in North Africa were stockaded and hunted for centuries—in the form of a canned hunt.*
Bottom: *In the seventh century BCE, lion hunts were the favourite sport of the Assyrian kings and took place mainly in their private hunting parks where they were carefully staged to put up a public display of power.*

record information about the traffic in exotic animals that we know had already been going on for centuries.

> In the yeare 1581 as king Phillip was at Lisbone, there was a Rhinoceros and an Elephant brought him out of India for a present, and he caused them both to be led with him into Madrid, where the Spanish Court is holden.

Likewise, when François Pyrard de Laval travelled across India between 1601 and 1611, he didn't mention the lion but was very clear about tigers:

> Of tigers there are a vast number in the Indies; indeed they are commoner than wolves here. It is a most ferocious and mischievous animal, which will not flee from men except they be in great numbers, but, on the contrary, will pursue, attack, and devour them.

Following in Laval's footsteps was Thomas Best who was in India from 1612 to 1614. He mainly focuses on his journeys across Gujarat but again does not talk of lions. Sir Thomas Roe, who visited Jahangir's court in the early seventeenth century as ambassador of James I of England, mentions a lion breaking into his quarters one night for some sheep in the courtyard;

he had to seek permission to hunt it as 'none but the king can hunt the lion'. What is of interest here is the tameness of the lion. Sir Thomas states unequivocally: 'Lions here are…feeble and cowardly.'

Notably, most travellers to the country only saw the lion at the courts of kings. Sir William Foster, who came to India in the seventeenth century, provides evidence of the status of India's lions as animals intended for the amusement of royalty:

> At night the Kings custome being to drinke, the Prince, perceiving his father to be merry, told him of this man. So the King commaunded him to be brought before him. Now while he was sent for, a wilde lyon was brought in, a very great one, strongly chained, and led by a dozen men and keepers; and while the King was viewing this lyon, the Pattan came in, at whose sight the Prince presently remembred his father. The King demanding of this Pattan whence he was, and of what parentage, and what valour was in him that he should demand so much wages, his answer was that the King should make tryal of him. That I will, saith the King; goe wrastle and buffet with this lyon. The Pattans answere was that this was a wild beast, and to goe barely upon him without weapon would be no triall of his manhood. The King, not regarding his speech, commanded him to buckle with the lion; who did so, wrastling and buffeting with the lyon a pretty while; and then the lyon, being loose from his keepers, but not from his chaines, got the poore man within his clawes and tore his body in many parts, and with his pawes tore the one halfe of his face so that this valiant man was killed by this wilde beast. The King, not yet contented, but desirous to see more sport, sent for ten men that were of his horse-men in pay, being that night on the watch; for it is the custome of all those that receive pay or living from the King to watch once a weeke, none excepted, if they be well and in the citie. These men, one after another, were to buffet with the lyon; who were all grievously wounded, and it cost three of them their lives.

> The King continued three moneths in this vaine when he was in his humors, for whose pleasure sake many men lost their lives and many were grievously wounded. So that ever after, untill my comming away, some fifteene young lyons were made tame and played one with another before the King, frisking betweene mens legs and no man hurt in a long time.

These menageries were not just a feature of the Mughal court—most rajas and nawabs in the continent had menageries as well as hunting parks. As a result, these provided a ready market for dealers in exotic animals. The business in imported animals would peak 200 years later but it was already flourishing; it needed to be if it was to keep menageries like those maintained by Emperor Jahangir well stocked. Thomas Coryate, a traveller who passed through the Mughal Empire in 1611, writes of this impressive collection:

> Of Elephants the King keepeth three thousand in his whole Kingdome at an unmeasureable charge, in feeding of whom, and his Lyons, and other Beasts, he spendeth an incredible masse of money, at the least ten thousand pounds sterling a day.

Sir William Foster corroborates Coryate's account and reveals that Jahangir's court and menagerie comprised 100 tame lions and 400 cheetahs. There were also 4,000 Persian horses, 6,000 Turkish horses, 12,000 elephants, 2,000 camels, 1,000 mules, 3,000 deer and buck, 400 hunting dogs and greyhounds, 500 buffaloes, 4,000 hawks and tens of thousands of birds, including pigeons. It is unsurprising that Jahangir had such an extensive menagerie given his reputation as a naturalist. He was a great experimentalist when it came to animals, especially exotics; in 1619, he even tried to cross-breed Barbari goats from the port city of Darkhar in Arabia with the markhor (*Capra falconeri*) and create a new species!

Going by travellers' accounts, Jahangir was probably the Mughal emperor most closely associated with the lion. For instance, *The Journal of John Jourdain* includes a wonderful account of the savage games the emperor played with his royal pets:

> The rest of the daie he employeth in seeing elaphannts to fight, and other sports. One of his sports is to bringe forth a wild lyon and lett him loose amonge the people, to see if there be any soe hardie as to stand against the lion; which if there bee, he is a man for him, and will doe him greate favor. As at one time he brought forth a lion amongst the Portugalls to see if anie would resist him, but they all ranne awaie except one; butt the lion cominge towards him, he went to defend himselfe as he might, and struggled a good while with him untill they gate both into the river... Yee, he awnswered, that there was more in him then in other men; which if it pleased His Majestie to employe him he should see itt. Whie, said he, wilt thou fight with a lyon? He awnswered that a lion was a beaste that had noe sensible understandinge, and therfore not fitt to be fought withall. Naye, said the Kinge, thou shalt fight with a lion; and therewith caused a lion to be brought forth and the man must fight with him hand to hand, onlie a gluffe on his hand, and a little trunchion of a foote and a half longe. Soe he fought with the lion a prettie space and overthrewe the lion; yett it bruised and tore the man soe with his clawes that hee died within a little space. These are some of the Kings sports.

While others have described lions at Jahangir's court, it is odd that the emperor's own memoirs do not mention a single lion. Some natural historians have explained this away by saying that this was because lions were commonplace but this contradicts everything we've seen thus far. In my view, the reason Jahangir does not dwell on lions or lion hunts is because the animal was so rare in his time as well as the epochs preceding it, as corroborated by other chroniclers and travellers in this period. In fact, over the centuries, accounts of encounters with wild lions actually keep dropping—hardly the sort of thing you would associate with a species that had successfully adapted to new territory.

Although some of the lions at court were used to fight men or other animals and even execute criminals, they were, by and large, expected to be royal accoutrements and for this reason needed to be properly tamed and domesticated. An early history of India gives a fascinating insight into how lions were tamed:

> For the lions to be safe for people they underwent special training and this amazing example reveals the process under which lions were trained in Awadh in the 17th century. When I arrived at Sidhpur on one of my journeys, I was encamped under two or three trees at one of the ends of a great open space near the town. [A] short time afterwards four or five lions appeared which they brought to train, and they told me it generally took five or six months, and they do it in this way. They tie the lions, at twelve paces distance from each other, by their hind feet, to a cord attached to a large wooden post firmly planted in the ground, and they have another about the neck which the lion-master holds in his hand. These posts are planted in a straight line, and upon another parallel one, from fifteen to twenty paces distant, they stretch another cord of the length of the space which the lions occupy, when arranged as above. These two cords which hold the lion fastened by his two hind feet, permit him to rush up to this long cord, which serves as a limit to those outside it, beyond which they ought not to venture to pass when harassing and irritating the lions by throwing small stones or little bits of wood at them. A number of people come to this spectacle, and when the provoked lion jumps towards the cord, he has another round his neck which the master holds in his hand, and with which he pulls him back. It is by this means that they accustom the lion by degrees to become tame with people, and on my arrival at Sidhpur I witnessed this spectacle without leaving my carriage.

Besides being a source of entertainment in palaces, lions were also hunted in a carefully managed way. In the Mughal era, the head huntsman was always close to the king and responsible for making his hunt a success. No effort was spared in safeguarding the king. Elephants were trained in advance on how to deal with tigers or lions.

The king would keep at least a hundred male elephants for himself, some of which would be trained to withstand weapons used in war as well as fireworks so that they were no longer afraid of them. The rest of them were trained for the hunts and taught not to be scared of lions and tigers. 'To teach them, they take a tiger-skin or lion-skin, and stuff it with straw. Then, just as if it were alive, they move it here and there by a rope. The driver encourages the elephant, and urges him towards the dummy, which with feet and trunk he tears to pieces.' The elephants were treated well and plied with spirits to encourage them to fight.

In the designated hunting areas, the Mughal emperors would wait till several hundred horsemen and soldiers drove beasts like lions, tigers or deer and gazelle towards the king's 'musket range'. When lion hunts are described, they tend to take place near tanks and inside fort walls:

> On the 4th the huntsmen sent news that they had marked down a lion in the neighbourhood of the Shakkar tank, which is inside the fort and one of the famous constructions of the rulers of Malwa. I at once mounted and went towards that game. When the lion appeared he charged the ahadis and the retinue and wounded ten or twelve of them. At last I finished his business with three shots from my gun, and removed his evil from the servants of God.

In his book, *Gods, Kings and Tigers: The Art of Kotah*, Stuart Cary Welch clearly states that slaying beasts like lions and tigers helped the ruler show that he protected his people against evil beasts, and the royal hunter was the embodiment of good as against his quarry, who was the embodiment of evil. Commenting on a 1770 painting of hunters slaying tigers, he states:

> Such scenes bring to mind ancient prototypes...Assyrian and Archemenian hunting reliefs, scenes painted for the Turkmans, Safavids and Mughals and Hindu portrayals of Gods vanquishing demons.

It was a symbolic hunt and must have also occurred at auspicious times to demonstrate the victory of good over evil.

I have examined scores of paintings of the Mughal school depicting lion hunts, and the first thing that struck me was that the painters seemed unfamiliar with the animal, depicting it in all manner of strange shapes and sizes, implying that the lion was rarely seen. On the other hand, Mughal representations of the tiger are far more realistic. The paintings are full of clues and serve as an excellent record of how royal lion hunts were stage-managed. For a start, the hunts do not seem to have occurred in a thick forest or in the untamed wilderness to be had in India. They appear to take place in designated hunting areas with forts and palaces in the background and even platforms to which animals are driven. Sometimes palanquins are shown waiting for the shoot to end to transport the hunters back to their palaces. Often soldiers surround the entire area, while accidental encounters are few and far between. The peripheries of the paintings depict small cities to show their proximity to the hunt. People walk around on the margins of the hunt. Nets around tigers and lions are visible. Moreover, if the paintings are examined carefully, long ditches surrounding the area add their artificiality to the scene. The Mughals were famous for fencing off roads for their hunts—these were used to drive and direct game. Fencing could be done over many miles and anywhere from 4,000–100,000 troops could accompany the hunt. All the Mughal miniatures I have observed point to lion hunts being artificial to the extreme. Groups of lions that appear young in age are surrounded by nets and driven towards kings comfortably ensconced in palaces or pavilions. Clearly, they must have been hand-reared and then released to be driven towards the spot where the ruler sat. The scenes appear almost comical, with lions surrounded by nets, armed guards or armies, like lambs awaiting slaughter. The thousands of beaters not only drove the animals into designated enclosures but kept them there till the last one was shot. All the royals seemed to do was to loll about on cushions or in pavilions and kill hundreds of animals that were literally driven to their feet. In some of these paintings, the public is also shown watching the massacre.

Perceptive travellers through Mughal lands show just how absurd these hunts could be. Writes Bernier:

A domestic elephant with its hind legs tied defends itself against a lion, probably in a lion enclosure in the nineteenth century.

As a preliminary step, an ass is tied near the spot where the gamekeepers have ascertained the lion retires. The wretched animal is soon devoured, and after so ample a meal the lion never seeks for other prey, but without molesting either oxen, sheep, or shepherds, goes in quest of water, and after quenching his thirst, returns to his former place of retirement. He sleeps until the next morning, when he finds and devours another ass, which the gamekeepers have brought to the same spot. In this way they contrive, during several days, to allure the lion and to attach him to one place; and when information is received of the King's approach, they fasten at the spot an ass where so many others have been sacrificed, down whose throat a large quantity of opium has been forced. This last meal is of course intended to produce a soporific effect upon the lion. The next operation is to spread, by means of the peasantry of the adjacent villages, large nets, made on purpose,

> which are gradually drawn closer, in the manner practised in hunting the nil-ghaux. Everything being in this state of preparation, the King appears on an elephant protected in places with thin plates of iron, and attended by the Grand Master of the Hunt, some Omrahs mounted on elephants, and a great number both of gourze-berdars on horseback and of gamekeepers on foot, armed with half-pikes. He immediately approaches the net on the outside, and fires at the lion with a large musketoon.
>
> The wounded animal makes a spring at the elephant, according to the invariable practice of lions, but is arrested by the net; and the King continues to discharge his musketoon, until the lion is at length killed.

What is vital to note here is the link between the king and the lion. According to Allsen, 'Hunting lions was…a mechanism by which a proper balance was struck between the menace of nature and its essential nurturing function. Because this was a vital spiritual and political matter, royal lion hunting became highly ritualized, that is to say, minutely stage-managed, even to the extent of breeding.' In his book, *The View of Hindoostan,* published in 1793, Thomas Pennant wrote that Akbar 'hunted the lion, the elephant, and the *yuz*, or hunting leopard, but more to show his imperial courage, and his skill in shooting with the fusil or bow, than from any pleasure he had in the discipline of the pack.'

It would have been impossible for anyone to question the courage of a man who fought a lion or a tiger, even in a controlled environment. Akbar is supposed to have, as a young man, killed a tiger on foot.

■

One of the justifications for these hunts, besides showcasing the power of the royals, was that it was good training for the army. 'In many parts of Northern and Central India imperial hunting preserves were established, and emperors, princes, and great amirs spent significant portions of many expeditions finding, riding down, shooting, and killing wild animals.' Hunting tigers was limited to the royals and only a selected number of amirs were allowed to participate in these hunts. 'Ordinarily, the great cats were shot by musket or bow and arrow from the backs of elephants but sometimes they were approached on horseback. Great men hunted elephants from the backs of other elephants.'

William Jardine adds more information on the Mughals in their royal parks and the way they were able to tame animals and hunt them:

> Beasts of prey were taken in this way and were kept in royal parks, to be hunted at leisure, or to be matched against each other at public fights.

François Bernier, who has left behind some of the best accounts of this period and provides many details of royal hunts, is categorical about the place of the lion in wild India. He writes in the mid-seventeenth century:

> Except in Kathiawar, lions are now never met with in any part of India.

This is exactly what Linschoten had said 100 years earlier—lions were not only rare but also extremely precious. Bernier describes the huge effort made to recapture an escaped lion: '[the] enraged animal leapt over the net, rushed upon a trooper whose horse he killed and then effected his escape for a long time. Being pursued by the huntsmen he was at length found and again enclosed in the nets.' The emperor's order must have been clear—no lion could be lost, even if recapturing it cost the army a few days.

After talking about how well-planned these hunts were—'[a lion] was at length found and enclosed in the nets'—Bernier writes that great care was taken to ensure that not a single lion was 'lost' because it is considered a favourable omen when the king kills a lion, and the escape of that animal is 'portentous of infinite evil to the state.' Even after a lion was killed, there were 'grave ceremonies'. The lion was brought before the general assembly of omrahs to be measured and minutely examined and every detail of the specifications was carefully recorded.

I would not be surprised if lion hunts were scheduled as

The keeper of the lions which lived in royal menageries across India and Persia (nineteenth century).

per astrological forecasts and even the time to slay it was predetermined. The huntsmen also told Bernier that feeding the lion opium to put it to sleep was not a must—just a full meal was enough!

■

As we get to the end of the seventeenth century, we find an account by the Neapolitan traveller, John Francis Gemelli Careri, that corroborates the observation made by the Dutchman Linschoten a century earlier. As he waits to leave Goa in 1696, he writes about going 'to see a lion brought by the viceroy from Mozambique, who was about to send it as a present to the Emperor of China.' It is quite clear that there were very few lions near the coast of Gujarat or Goa at the time, which is why they were being imported from Africa. In fact, these coastlines were full of tigers. Careri talks of countless tigers that were hunted near Daman and Diu—adjacent to present-day Gir, which supposedly is lion territory—but he does not mention lions. So where were

These lions are completely stockaded before being killed mercilessly. Here, eight young lions have been shown and the scene typifies the practice of 'canned' hunting by kings and noblemen.

the lions of Kathiawar and India in the seventeenth century? Was the presence of the lion in Kathiawar at this time just a myth?

Meanwhile, this is one of the first mentions of the flourishing trade in exotic animals—especially lions—that we will look at in greater detail in the next chapter. What is clear from the accounts of Careri and others is that there were hundreds of ships voyaging from Africa (especially Mozambique) to India. There were Dutch, British and Portuguese vessels, not to mention Mughal fleets, carrying as part of their cargoes, lions, cheetahs, zebras, giraffes, elephants, dogs and horses along with goat, oxen and buffaloes to feed them and slaves to take care of them.

Although this is a subject that deserves to be extensively researched and written about, Sidis, the tribespeople of Abyssinian descent who live in Gujarat, are without a doubt descended from Africans who served in the thousands of ships that docked in Gujarat every year for centuries. Some historians believe they arrived in Bharuch as early as the seventh century. Upon arriving on these shores, they were employed by royals and wealthy merchants as bodyguards, food tasters, soldiers, servants and animal keepers. Some of them became traders, and many even grew powerful in the courts of kings. In 1490, when Firoz Shah Tughlaq, one of the kings of the Delhi Sultanate, was succeeded by his son, real power at court vested in the hands of an Abyssinian slave, Habash Khan; Khan was killed by another Abyssinian, Sidi Badr Diwana, who then seized the throne. Another powerful royal of African descent was the Habshi ruler, Sidi Yaqut, who was given command of the island fort of Janjira by the ruler of Ahmadnagar in the sixteenth century. For the next 200 years, Yaqut's descendants were the 'unchallenged masters of the coasts, whether serving Ahmadnagar, the Mughals or their own interests.' It was only in 1819 that control of this territory passed to the British.

African immigrants and slaves seem to have been especially skilful at handling animals. A Mughal miniature painting depicts an African leading a cheetah by a leash in the court of Akbar at the time of the birth of his son, Murad, in 1570. Whether to look after animals or to tend to their masters, thousands of African slaves entered India each year and merged with the locals.

Indeed, they have always been considered to be experts on understanding the behaviour of lions. They knew the territories of individual lions and could drive them right up to the hunter's perch. Today, their villages are scattered across Gir and I have spent many an evening with them listening to African drums. Most of them, to this day, consider themselves to have directly descended from the original lion trackers of the Nawab of Junagadh and other local kings and rulers. Helene Basu, professor of Social Anthropology at Westfälische Wilhelms-Universität Münster, and a leading scholar on the Sidis, states that:

> There is a widespread stereotype that associates Africa with wilderness and lions. In fact, when the nawab brought African lions to the Gir forest he also settled some Sidis to care for them.

It would be appropriate to close this chapter on lions in the period before the British Raj with a quote from the seventeenth century traveller, John Francis Gemelli Careri, which ties together the presence of Africans in India, the dog-like lions to be encountered in Gujarat, and the ease with which these tame beasts could be killed. He writes: 'There are [an] abundance of Cafres and Blacks… These slaves are carried to sell at Goa [and other] Portuguese towns by the company's ships.'

Careri then narrates how they hunt lions:

> They kill Lions for sport; for when they see one astray in the Woods, one of them advances with two small Cudgels in his Hand, and clapping one of them into the Lions Paw, plays with the other: In the mean[time] while the next Black to him very dexterously takes the Beast by the Testicles, and then they beat him to death. So when they would have a Lion quit a Cow he has seiz'd, they draw near, and saluting him after the same manner as is us'd in Africk, to Persons of the greatest Note; that is, lying down on their side, holding up one Foot, and at the same time making a Noise with Hands and Mouth. This was generally told to me by the Portugueses; the Reader may believe what he pleases.

The lions here appear to have been bred in hunting grounds as they seem tame enough to be jumped on and beaten to death. I wonder whether these descriptions from seventeenth-century India reflect the connections between the Africans and their lions as they existed in Africa at that time. Did the Masai behave in much the same way when they hunted their lions?

Some of the lion art, to say the least, was bizarre. All over the world, lion art spread as human beings travelled from Mexico to China. Lions were captured across the world and bred in countries that had never known wild lions.

Next page:
The royal lion and the crown, integrally linked to kings and emperors.

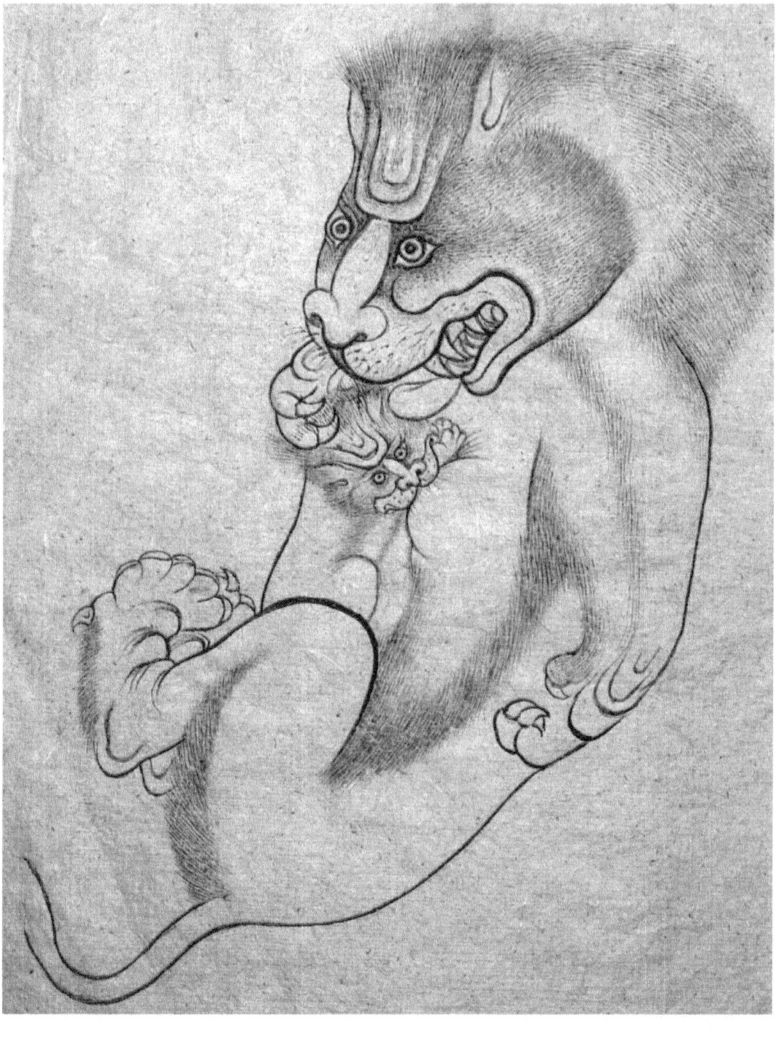

THE LAST LIONS

by VALMIK THAPAR

Between the late-eighteenth and the early-nineteenth century, as Mughal power began to fade, the British went about consolidating their power in the Indian subcontinent through a mixture of opportunism, deceit and pitched battles with a variety of opponents. For relaxation, among other leisure pursuits, they hunted. A variety of British 'sportsmen' have left behind accounts of hunts, and it is fascinating to read through them to tease out their encounters with the lion. While there are plenty of lion 'stories', there are hardly any eyewitness accounts of the animal in the wild. The number of lions bagged was tiny when compared to other wildlife and only confirms the trend we first noticed in pre-Islamic observations about the lion as well as accounts of the animal in Islamic and Mughal times. What is critical to note about British accounts of India's wildlife is that, unlike the Mughals, who limited themselves to stage-managed hunts, the British actually ventured into wild India in pursuit of game or as administrators and explorers, as a result of which they succeeded in painting a much more accurate picture of the situation on the ground, and not just record wildlife in royal hunting preserves or game parks.

Late in the eighteenth century, Captain Thomas Williamson, who published a monumental record of hunting and wildlife related activity of the British in the country, records what seems to be yet another import: 'As to lions there are none in Hindoostan… The only one ever seen in that country was that sent from Ghod in 1781, as a present to Mr. Hastings, then Governor General of India. It was considered as a unique animal and had been brought from the north of Persia, where it's said to abound.'

His compatriot, Thomas Pennant, underlined Williamson's observation in his book, *The View of Hindoostan* (published in 1793), wherein he states: 'Whether they exist at present is doubtful, being animals at least very rare at this time.' Pennant believed that in the time of Jahangir they must been plentiful, although as we have seen, the evidence points to lions being bred in hunting parks and released for pre-arranged royal hunts.

> In the neighbourhood of these forests, and that of Rhotas Gur, are numbers of lions. Those who deny that those animals were natives of India assert that here was a royal menagerie, and that the breed was propagated from the beasts, which had escaped. The Ayeen Akberry, ii. 296, relates many instances of the valour of Akbar, the Great, in his engagements with this tremendous animal, but is silent whether they had or had not been aborigines of Hindoostan.

And so, once more, serious doubts emerged about lions being 'aborigines of Hindoostan', with many speculating that Indian lions (in the wild) were originally escapees from royal menageries. The debate raged on because there were no lions to be seen, the numbers never added up and it was much the same as in the sixteenth century.

In eighteenth century India, there were probably around 400-500 menageries ranging from little ones in small kingdoms like Bhavnagar to large ones in states like Mysore, Udaipur and Hyderabad which had tigers, lions and even rhinos. In maintaining such menageries and hunting parks, the maharajas and princes were continuing a trend started by the Mughals, and the empires that succeeded them. The lions were symbolic of the king and hunting them, even in controlled conditions, was a vital initiation from prince to king. The emperor and his

The lion and the elephant—a typical confrontation in the private hunting grounds that were scattered across India.

family's prerogative—killing a lion—meant the power to rule effectively. Shah Jahan killed one with a sword, as did Aurangzeb's son, Mahmud.

Allsen writes about a game reserve that existed during the time of Akbar along the Narmada River, which measured between 96 and 128 square miles in area, and I think it is more than a coincidence that one of the earliest and most detailed eighteenth-century accounts of the Indian lion describes it as being found right where a historic Mughal game preserve existed. In his memoirs, James Forbes writes about a hunting expedition Sir Charles Malet, then President of Cambay, undertook in Gujarat in 1781. The hunters were looking for tigers and never expected to find lions, indicating that they were so rare in the area that most Europeans were unaware of them.

> Having watched for two nights in vain for tigers, on the third evening we tied lures of goats and asses under the trees, in three different places, and at each of these stations three marksmen, including myself, watched in a tree. About midnight, four animals, which we imagined to be tigers, but afterwards discovered to be lions, having at some distance

A lion in India ambushed by armed men on foot. Finding a lion in the wild was always a surprise.

taken a momentary survey of the goat tied at one of the posts rushed furiously on it; and the largest of them seizing it by the neck, with one shake broke the bone, and the animal was instantly deprived of life. The lion then made an effort to carry off his prey, which being purposely bound with strong cords he failed in the attempt. At that instant two of the marksmen posted with me in the tree fired and wounded him, but he suffered only a momentary stupefaction, for immediately recovering, he quitted the slain goat and retired. One of a smaller size instantly came forward and seized the goat, when the third marksman fired, and wounded him; he also directly retired; but, by the light of the moon, we perceived that they both retreated with difficulty.

This beast was called by the country people oontia-baug, or camel-tiger, and is by them esteemed to be the fiercest and most powerful of that race. His colour was that of a camel, verging to yellow, but without spots or stripes; not high in stature, but powerfully massive, with a head and foreparts of admirable size and strength. He was killed near the village of Coora, on the banks of the Sabermatty, fifteen coss from Cambay.

At the end of the eighteenth century, Forbes was convinced that lions existed only on the borders of Persia. The tradition of rearing wild animals for sport and entertainment continued into the early and mid-nineteenth century before the 1857 Uprising. This enormous convulsion within the country, which finally brought the British to absolute power in the subcontinent, had an impact on every aspect of the country, including the menageries of princes and nawabs. But in the first half of the nineteenth century, before war and civil unrest distracted Indian royals, hunting and the accumulation of exotic animals was among their main obsessions. Several animal traders ensured their menageries were well stocked and hundreds of vessels traversed the waters between India, Africa and Persia. I believe that Gujarat was the region on the western coast where animals like lions and cheetahs that were imported were acclimatized.

In some of the biggest menageries, lions and tigers were plentiful. The following is an excellent description of one such menagerie belonging to the Raja of Coorg:

> On a signal given by the Rajah, a folding door was thrown open on one side of the court, and in stalked two immense royal tigers, held by several men on each side by long but slight ropes attached to collars round the animals' necks. These beasts appeared very tractable, for they allowed themselves to be led close to us. I confess I did not much like this degree of propinquity, and eyed the slender cordage with some professional anxiety. Meanwhile the Rajah and his son, and the officers of the household, appeared quite unconcerned, though the tigers passed within a few yards of them, and, as it seemed to me, might easily have broken loose. What degree of training these animals had undergone, I know not; but, after a little while, the Rajah, probably to increase the surprise of his guest, directed the men to let go the ropes and to fall back. There we sat, in the midst of the open court, with a couple of full-sized tigers in our company, and nothing on earth to prevent their munching us all up! The well-fed and well-bred beasts, however,

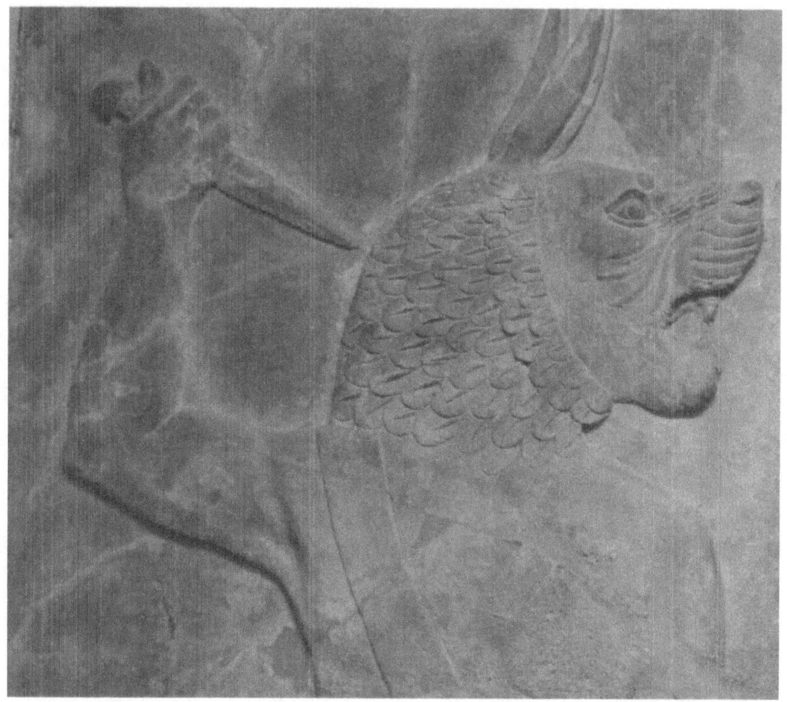

This is a depiction of a lion-headed human being (protective spirit) from the Assyrian period of the seventh century BCE. Such themes travelled across world as trade grew; even in India's history, there is the myth of Narasimha who was half lion, half man.

> merely lounged about, rubbed their noses together, and then tumbling on the ground, rolled about like a couple of kittens at play. I could, however, detect the Rajah spying at me out of the corner of his eye, and half-smiling at the success of his trick. After a time the men were recalled and the tigers dragged off…
>
> A pair of lionesses and two furious-looking buffaloes were then introduced…

What is fascinating about the lion-buffalo fights is that African buffaloes were imported as they were wild and fierce enough to fight with the lions! Were the lions African? Whichever way you look at it, exotic animals were brought in from across the seas, be it from Mombassa, Zanzibar or Mozambique and all of them were dropped off on the western coast. Gujarat may have been a big landing ground for exotic animals to acclimatize themselves before they were sent to different destinations.

Lucknow was another kingdom that possessed enormous menageries possessing many lions.

> Organizing fights between beasts of prey in Lucknow was confined to the royal family and nobles of the court. Keeping the animals, training them, controlling them after the fight and protecting spectators from injury were beyond the means of the richest men, let alone the impecunious. For this reason fights between beasts of prey were only witnessed in Lucknow whilst the court existed, and when the court disappeared so did the terrifying amphitheatres...

Baroda had large menageries including a large collection of lions kept in enormous cages. It is clear then that the presence of the lion from the late-eighteenth century all the way through the middle of the nineteenth century followed the pattern established in previous centuries—it was rare, usually found in the vicinity of ancient hunting parks and preserves, and was a vital part of the menageries maintained by the rajas and nawabs. It was these menageries that could facilitate the release of lions for a hunt. One of the rarest and most unique encounters with the lions around Baroda comes from a group that thought they were after tigers. This was also a time when the existence of lions in India was being questioned.

The debate about the status of Indian lions seems to have continued from the end of the eighteenth century to the beginning of the nineteenth century because they were so rarely sighted. In the March 1817 issue of the *Asiatic Journal*, there is a description of a lion hunting party at Baroda on 26 June 1816 where the author states, 'indeed the existence of this species in India has been questioned…' He describes the shooting of two lions in a garden located a mile to the west of the city. The party were expecting tigers (also rare in that area) but found lions instead—an animal that had never before been seen there. Describing the place where these two lions were found, the

Englishman says: 'The place in which they were said to have taken shelter was covered by bushes of the Mogri flower plant, extremely thick, and standing about four feet high, with narrow pathways occasionally intersected by hedges of the prickly milk bush, and low and thick ramifications of the aloe tree.' The two animals were young lions and were shot as soon as they got within a few paces of the party, who then, 'with the bayonet and swords put an end to [them].' What is even more amazing is 'on being brought to the Residency and inspected, these animals were sent to his Highness Futteh Singh at his own request.' This was probably done as the lions were the personal property of the king and, even in death, had to be returned to him.

As far as I am concerned, these were escapees from the royal menagerie of Baroda and this early description typifies the kind of docile lions that were found in the 'wild'. In fact, hunting records through most of the nineteenth century attest to the scarcity of the lion. M. A. Rashid, a senior officer of the Indian Forest Service, has kept a record of lions killed in India from 1781 to 1891 and the total comes to the unbelievably low one of 50-60 animals shot over a 110 year period. (This does not include Colonel Smith, of whom more later.)

Observers like R. G. Burton—a prolific shikari and writer—are surprised by the absence of lions. According to him, 'It is curious that 30 years after the last lion was killed in Hariana no traditions of the animal remained in that part of the country.' If only Burton had believed that the lions killed in Hariana (present-day Haryana) were escapees from menageries, I believe he wouldn't have been so puzzled. Hunting records that Burton unearthed testify to the meagre number of lions shot:

> 1847 – one in Damoh; 1850 – 2 in Guna; 1860's – 11 in Guna; 1866 – 1 in Allahabad; 1871 – 3 in Gwalior; 1872 – 1 in Guna.

Burton's secondary records show 19 lions shot in the period between 1847 and 1872, most of them in Gwalior and Guna, which not only boasted huge menageries but also restocked their forests with African lions.

> There were Lions in Gwalior territory in Central India long after this date. The late Colonel Cunliffe Martin, of the Central India Horse, and formerly of the 14th Light Dragoons, with which he served through Sir Hugh Rose's campaign in 1858, and the late General Sir Montague Gerard, both told me that they had shot a number of Lions in the neighbourhood of Goona. Gerard shot one on Waterloo Day 1872, and the last one shot appears to have been killed by Colonel Hill in the following year; but Hughes-Buller of the Central India Horse saw one in Gwalior State in 1884.

In Gujarat—exclusive of Kathiawar—the last survivor is said to have been killed in 1888. Since that date, there have been no lions spotted anywhere other than in the Gir forest.

Captain Mundy, quoted by Brig-Gen R. G. Burton, reports:

> Three shikaris arrived from Colonel Skinner to assist us to find a Lion between that place and Hansi...the royal race of the forest—like other Indian dynasties—is either totally extinct, or has been driven farther back into the desert.

Seeing lions was so infrequent that it was not uncommon for narratives to bemoan their absence. Victor Jacquemont, a nineteenth-century traveller to India, writes:

> My last letter, my dear friend, was addressed to you from Kithul, in the country of the Seikhs, dated the 2nd March. For a fortnight I was running after lions there, which we went in search of nearly to the edge of the desert of Bikaneer, but we did not so much as catch sight of one. In this short space of time, deducting the lions, I saw more of the East than in the whole year which had elapsed since my arrival in India.

If the British could not find lions, neither, it seems, could the locals. Even shikaris were unfamiliar with the beast, which would certainly not have been the case had it been a prevalent indigenous species. An account of the shooting of a lion appeared

The lion under the tree with the sage's blessings. Tame lions were probably kept for release on special occasions to facilitate hunts.

in *Oriental Sporting Magazine,* published in the 1870s, in which W. Kelsey, a contributor, writes:

> The sun had already set when they were astonished to see a fine lion striding towards them with a most majestic air. He was, however, soon killed with a few shots... The astonishment of the natives was even greater than that of the sportsmen for they had never seen such an animal.

I think what is critical to note here is the 'astonishment' of the 'natives'. The incident took place in the nineteenth century, and had the lion been a native predator, it is very unlikely they would have been astonished.

Other aspects that betray a lack of knowledge of the lion as a native species can be traced to things like the creation of a series of paintings of lion hunts in Bundi. There were probably large animal menageries in Kotah and Bundi, and it is likely that lions were released into the forests of that kingdom as a result of the numerous battles between the British and Indian rulers at the time. Apparently, Bishan Singh, the ruler of Bundi between 1804 and 1821, ordered his court painter to create a series of paintings to commemorate the slaughter of 100 lions!

This sounds highly improbable. The Bundi forests, which are contiguous to Ranthambore, are packed with tigers and the two could not have possibly co-existed. Moreover, the paintings which have been studied by Jiwan Sodhi in his book, *A Study of Bundi School of Paintings,* often show unfamiliarity with the lion.

▪

Throughout the period we're talking of, there are numerous accounts of hunts by travellers like James Tod (whose book, *Annals and Antiquities of Rajasthan*, is now a classic) and others but virtually none of them mentions lions as being a part of the prize.

Sporadic mentions of lions being killed only serve to underline how rare they were. In *The Bengal Sporting Magazine* of October 1838, a writer signing himself 'Kattywar' stated that he shot 11 lions between the fourteenth and twenty-fourth of May, to which three more were added between the nineteenth and twentieth of July. Throughout the later volumes of *The Bengal Sporting Magazine, The India Sporting Review* and the revived *Oriental Sporting Magazine* which followed it, there are occasional articles that describe the shooting of lions. In *The Asian,* dated 30 June 1885, Colonel Martin relates how he and General Travers killed two lions on a hill to the west of Goona in Gwalior in 1860; two years later, he and Colonel Beadon at Patulghur, some 70 miles northwest of Goona, bagged no less than eight.

The Bengal Sporting Magazine of 1833 contains an account of a lion in Haryana that was killed in a fairly 'tame' way. Another example of lion hunts at the time is the following account by the sportsman and naturalist, Layard Livingstone Fenton:

> In spite of the fact that a certain amount of protection is accorded by the Junagadh Darbar to the Indian lion, which at the present day is only to be met with in the Gir Forest in Kathiawar, there is no doubt, I am afraid, that it is gradually but surely approaching extinction.
>
> The Late Sir Edward Bradford, informed me that when he was serving in the Central India Horse, 26 lions were shot

Shooting lions in private hunting grounds was strictly controlled and meant for those that were most important—especially princes and kings who had never shot a lion before. It was like an initiation rite for the prince or king.

> in the neighbourhoods of Goona and Saugor by the officers of his regiment and I have seen it recorded…that in the year 1832 the officers of the 23rd Bombay Cavalry used to hunt lions on horseback from the now deserted station of Deesa in Guzerat. They have, however, long since disappeared from all these localities, and the last lion killed outside Kathiawar limits was one which was shot by the late Colonel Heyland, of the old 1st Bombay Cavalry, on the Deesa race-course.

What is interesting to note in the above account is the use of the phrase, 'the last lion'. As we have seen earlier, it has been used repeatedly over various centuries by unrelated witnesses, just as the observation that there were 'no lions in Hindoostan' is encountered over and over in records. To my mind, this is not just a coincidence but a very accurate and specific insight into the fact the lion population in India must have been tiny—that is, it wasn't shot out of existence, but had always been rare, which in turn can only mean it was never an indigenous species that had established itself in this country.

▪

To continue examining evidence from hunting records, according to Paul Joslin, a senior wildlife scientist, 37 lions were shot between 1863 and 1888 which is a miniscule amount when compared to the 35 to 50,000 tigers that were shot at an average of 1,600 to 2,000 each year and the 70 to 90,000 leopards that were shot during the same period. Mahesh Rangarajan believes that between 1875 to 1925, 80,000 tigers and 150,000 leopards were shot. In the same period, barely 30-40 lions were shot, including those from the intensively managed Junagadh population. Adding up all the reports of lions shot in the nineteenth century including many figures that serious observers believe are exaggerated, it is difficult to reach a figure of 500 even if you are being generous—a number that amounts to less than two a year. In the same period, a conservative estimate of the number of tigers shot would be 150,000.

Even those accounts—unsupported by any real evidence—which thought the lion had once been numerous are puzzled by its absence in the nineteenth century. In 1831, Major Brown, an officer of the British army, writes:

> The lion was once very numerous in Haryana but not one [is] to be found.

The naturalist, Edward Blyth, writing in *The India Sporting Review* in 1856, states of lions:

> It is curious that not even a tradition remains of the former existence of that grand and most prominently conspicuous animal in the Hariana territory.

Another observer in the nineteenth century, naturalist Arthur Coke Burnell states: 'Lions (*Felis leo*) were formerly found in the North of India pretty generally, as the trite allusions to this animal in the Sanskrit books prove; it is now very uncommon in most parts of India.'

A contemporary naturalist like Divyabhanusinh postulates that it is possible that the species may have been present in one part of the country, and may have 'drifted' to another part—that is, Gwalior's lions may have drifted to Kotah (now Kota) in the 1930s. When a lion was spotted in the jungles of Kotah adjoining Gwalior: 'The shikaris of Kotah who had not seen a lion and were not aware of Kotah's past history, reported to the palace "that a wild animal like a tiger resembling the colour of an ass was seen in the forest" according to maharao Brijraj Singh of Kotah.'

I disagree with Divyabhanusinh as, even if a local shikari had not seen lions in his lifetime, it would mean that the oral traditions of this animal had never been passed on which was obviously not the case. It is clear that the locals never saw lions in the region, even though Divyabhanusinh also mentions that 'ten lions were shot in Kota, Rajasthan in 1866.' This entire incident goes to show that there was good reason to believe these were captive lions released for the hunt as opposed to being indigenous to the area. We know that the kings of Kotah shot in a private hunting ground called Alnia and many of the lion hunts depicted in paintings show lions being slaughtered in a stockade. If you examine such scenes, they are totally unnatural and reaffirm the theory of the canned hunt.

▪

Other clues that establish the scarcity of lions and the strangeness of their provenance are also found in accounts of hunting expeditions. Captain William Cornwallis Harris (1807-1848), for example, talks of trying to shoot lions but wonders why none of the lions that were bagged had reached adulthood. 'They are young and...[it is presumed that] all the adults have been shot.' This assertion, which pertained to a hunting expedition in 1838, is as absurd as the claim that Akbar had caught 1,000 cheetahs. Had the British really managed to shoot every adult lion in India? If so, how many adults had there been? What is more likely is that lions were generally scarce; a few were probably feral and adults only reappeared when they were brought to the country as imports from Africa by princes and maharajas.

▪

Besides the general scarcity of lions during this period, what is interesting is that, in many of the lion encounters, the animal appears to be unusually tame, or at any rate not fierce, which is not the case with the species in Africa. A comment by a correspondent of *The India Sporting Review* in the 1850s quoted by Burton about his hunting exploits makes that point:

> They did not always show fight but generally slunk away. When they did fight, they took to an open plain. It was difficult to get an elephant to face them and few would stand the charge.

These lions must have been raised and then maintained in a hunting park or they must have escaped, gone feral and survived on livestock. Again, a certain Captain Outram, the head of the Bhil Regiment, had an encounter with a lioness and her cubs in Kathiawar in the early nineteenth century which would have been almost impossible to contemplate with truly wild lions:

> While on the subject of hog-hunting, I cannot refrain from mentioning an adventure of Outram's, when engaged in that sport in Kattiwar. Two or three of his brother officers were with him, and, having been disappointed in the hog they had news of, they separated, walking their horses across a plain, ready for anything that might turn up.
>
> Suddenly Outram spied a couple of creatures that looked like kittens playing under a bush of prickly pear. They turned out to be lion cubs. Jumping off his horse, he caught them and stuffed one into each pocket of his coat.
>
> When in the act of remounting his horse, he heard a roar, and saw the lioness coming right at him. Springing into the saddle, he put the spurs into his horse, and dashed across the plain, closely followed by the mother of his captives.
>
> Before he could get the speed up, she gained rapidly upon him, and gathered herself up for one tremendous bound. She missed her mark by about a foot, but her claws actually combed the horse's tail. Before she could repeat her spring, Outram and the cubs got clear away. Outram, when fleeing

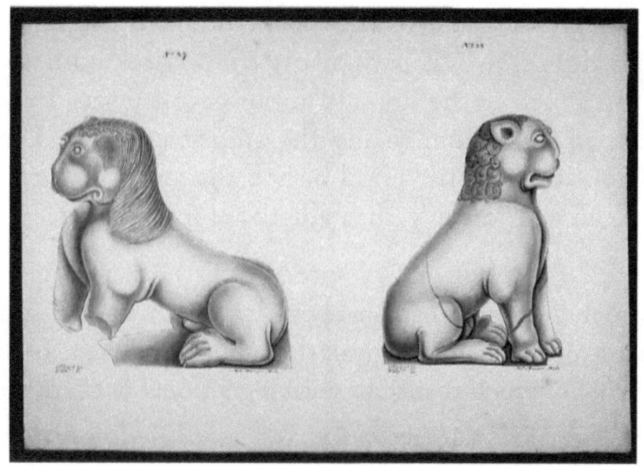

Guardian lions from the great stupa of Amaravati. These guardians reflect the alien lion art that was frequently seen across India.

> for his life, saw one of his companions at a distance, and, thinking he might as well share the excitement rode right for him. His friend, who was not a particularly bold horseman, no sooner saw the nature of the chase, than he pushed his horse up the rugged side of an adjoining height, and never looked behind him till he reached the top.
>
> He was not a little proud of the exploit, and would afterwards point to the steep as a place he rode up when hunting, carefully omitting all mention of the lioness.

I believe that the lioness's naturally ferocious instincts may have been dulled by the fact that she was descended from domesticated lions.

Before I conclude this account of the lion in the late-eighteenth and the first half of the nineteenth century, it would be pertinent to take into account the observations of Bishop Heber, who had come to lead the Christian church in Calcutta and proved to be an unusually clear-eyed observer of the India of the time. He writes that the depictions of lions he saw carved on the lion gate of the Jagannath Puri temple seemed to have been taken from the signs to be found on inns in England. It is a fascinating observation and as it was quite common for depictions of exotic animals in the world of art to travel far beyond their places of origin (this was particularly true of the lion), it could be argued that much of the lion art in our country was derived from elsewhere, and not from seeing the animal up close. There were other observations of the good bishop that are worth recounting. He states, after a journey through the upper provinces of India, that:

> The Sahibs from Shahjehanpoor…[said they] had hunted and killed the tyger. They call this animal not 'bagh' or 'bahr' but 'Shehr' which is strictly speaking a lion: but there are no lions in this part of India...

The locals told Heber that for them, *shehr* and *bahr* were the same animals. Burton expands on this theme:

> It is difficult to distinguish in Oriental records the Lion from the Tiger. Persian was the language of the Memoirs, as it is still the Court language of India. In Persian the Lion was sher and the Tiger babr, Jahangir refers to an animal he killed near Giri as sher babr, which seems rather to specify the Tiger.

And it was not just Bishop Heber who was surprised by the strangeness of lion art in India, especially in Puri. Edward Archer noted it too:

> The eastern or lion door [at Puri] is the principal entrance and is guarded by two stone animals, which the most depraved imagination has denominated lions: but they are as like whales as they are to that noble animal.

■

During the time immediately preceding the Uprising of 1857 and its aftermath, the subcontinent was ravaged by violent conflict, as the British fought to bring those who had rebelled against them under their control. The violence and disruptions of the time naturally affected the animal population as well. As the princes and maharajas went into battle, the upkeep and protection of their menageries and hunting parks naturally took a backseat, and scores of animals must have escaped. We have seen that Lucknow's menageries were among the best-stocked with lions and other exotics. During the hostilities, hundreds of lions must have escaped their incarceration, and while allowing for exaggeration, this is probably the only reason why sportsmen like a certain Colonel A. Smith could have told fanciful stories about his bag of lions.

Colonel Smith claimed to have shot 50-300 lions around Delhi and Haryana during the troubled period before and after the Uprising—and this is where not one lion was found in 1931 by Major Brown! Most serious observers regard Smith's record as a figment of his imagination, but it is likely that during this period of unrest, many menageries were looted and scores of lions fled and that Smith may have succeeded in killing most of this tame or captive population. Some may have even been killed in the private hunting parks of the local kings. There was a definite link between bloody conflict and the death of lions.

After the 1857 Uprising was quelled and the British Raj consolidated its hold on the subcontinent, we find accounts of some sightings of lions, animals being shot that may have been escapees, as well as the resumption of the breeding of lions and imports from Africa by nawabs and princes who resumed their tradition of hunting and breeding.

In the years following the Uprising, animal traders flourished. One of them, Edward Blyth, bought up the menagerie of the Nawab of Awadh after he was deposed. Blyth and others of his ilk found that rajas and maharajas, freed from the distractions of war, proved to be good customers. By 1859, Blyth was dealing with most of them, although Indian royals were not the only customers for animal traders like him.

> Blyth was part of the chain of which Bartlett and Jamrach's formed the intermediary links: from him, and from other colonial officers, flowed a stream of novelties on which the Zoological Society and the Natural History Museum depended.

Through these established trade links, animals were exchanged on both sides—wild animals were brought to Calcutta and unloaded in great numbers, from where they were sold to 'the lower-class Europeans in India [who] were fond of such creatures as pets.' Upper-class Indian society, too, enjoyed indulging in such novelties from other nations, '[just] as the British market was enthusiastic for Indian creatures.'

Perhaps the most famous importers and exporters of exotic animals of the time was the colourful Hagenbeck family, which had a presence in Africa, Germany, Sri Lanka and India. Nigel Rothfels writes:

> The beginning of the Hagenbeck exotic animal business can be traced to March of 1848...with almost every year the shipments of animals grew in frequency and size.
>
> When we consider the sheer volume of animals moved by the firm from the 1860s to 1880s, it becomes clear how Hagenbeck rose to such wealth and fame. By the mid-1880s, for example, Heinrich Leutemann noted that, by Hagenbeck's estimate, in the first twenty years of the company he had sold at least a thousand lions, three to four hundred tigers, six to seven hundred leopards, a thousand bears of different varieties (at one time he had forty-two at the same time), and around eight hundred hyenas. Some three hundred elephants had passed through his hands, of which sixty-three Indian and four African elephants had come in 1884 alone.
>
> About lions he [Hagenbeck] says, 'Without exception, lions are captured as cubs after the mother has been killed; the same happens with tigers, because these animals, when caught as adults in such things as traps and pits, are too powerful and untamable, and usually die while resisting.'

By the 1890s, Hagenbeck was selling hundreds of game animals across the world. It is evident he had strong links with the subcontinent because of his half-brother, John, who had settled in Colombo by the end of the 1880s, and his roaring trade in tigers.

Peter Ryhiner was another animal trader who spent months in India catching dozens of tigers near Mysore and lots of elephants and rhinos in the East—136 Indian elephants, 62 tigers, 23 black leopards and thousands of monkeys were a small part of his Indian efforts. His business was spread over Africa and South America. Frank Buck was doing much the same thing as he brought American bison in and took elephants out. The art of trading in live animals was at its peak in the first part of the twentieth century. Ships left India stuffed with native species, and others brought in a host of exotic animals.

African records of hunting and capture show that, between 1903 and 1911, 256 tons of ivory (representing 1,200-1,500 elephants each year) and 53 tons of rhino horn (representing 2,000 to 2,300 rhinos each year) and at least 1,000 wild animals were exported from Tanzania alone.

▪

I believe Indian royalty continued to import, tame and release lions into their game parks from which some inevitably escaped and their apparent tameness continued to astound observers. There was no great mystery to this in hindsight, but at the time it continued to occasion amazement.

In his book, *Sterndale's Mammalia of India*, written at the end of the nineteenth century, Sir Robert Armitage Sterndale quotes Bishop Heber of a century ago on lions that were shot:

> Those who have been killed in India instead of running away when pursued in a jungle, seldom seem to think its cover necessary at all. When they see their enemies approaching they spring out to meet them open mouthed, in the plain like the boldest of all animals – a mastiff dog. They are generally shot with very little trouble…

Compare this creature with the Algerian lion that was a truly wild species, which had almost become extinct by the mid-to-late nineteenth century. Both shared a North African point of origin and although superficially the Indian lion may have differed from the African one because of its lack of a mane, contemporary observers like Edwin Ward considered 'the Indian and African species identical, and the mane entirely a question of climate; just as tigers from cold regions, such as Amoor, are adorned with a ruff denied to their brothers in hotter latitudes.'

But the similarities seem to end there. Writes Jules Gerard about his encounter with a pride of Algerian lions in the nineteenth century in Africa:

> A family of lions...was coming down the path towards the ford, which I commanded... They stopped at fifteen paces from the bank of the river and [watched] me with looks of astonishment. According to my plan of attack, I aimed at the shoulder of the foremost and fired...The whole party pressed forward to see the lion, and to fire at him all at once. Unhappily it is too late. The lion, finding he is discovered, falls on the band, crunches the head of one, deprives another of his eye, lacerates the shoulder of a third, and disappears with a bound back into the wood, without giving time for a shot to be fired.
>
> Then the most deafening cries are raised; such a hubbub is produced that nothing can be heard; everyone accuses his neighbour of being the cause of the catastrophe...indeed the lion, irritated at the noise, and excited by the blood which he has just shed, returns across the wood, roaring, breaking, upsetting everything in his way, and with uplifted head and open mouth falls upon the band, who this time are not taken by surprise, and receive him with their shots fired point blank.
>
> The lion, riddled with balls, drops down in the middle of the party, and seizes with his mouth and claws everything within reach, biting and tearing away until he expires from his previous wounds, or receives one more bullet for his coup-de-grace.

None of the encounters with Indian lions that I have researched show the animal displaying behaviour that is even remotely similar to the courage and ferocity displayed by the Algerian lion. Indeed, the lions in India seemed to have been defined by their

The locals bow to the power of the Algerian lion which became extinct in the nineteenth century.

tameness and their inability to behave in the wild as any wild predator would.

Lions were seen in near-domesticated situations around tents and even bordering small towns; in 1886, Layard Livingstone Fenton, saw one trying to enter one of his tents. Few considered at this point that the lion was important as a part of India's natural world as they were seldom seen by anyone.

▪

By the end of the nineteenth century, what remained of the species was evidently declining:

> The lion is now confined to the Gir, or rocky hill-desert and forest of Kathiawar. A peculiar variety is there found, marked by the almost total absence of a mane; but whether this variety deserves to be classed as a distinct species, naturalists have not yet determined. The lion has now almost entirely disappeared; and the official Gazetteer of Kathiawar stated that there were in 1884 probably not more than ten or a dozen lionesses left in the whole Gir forest tract. They are strictly preserved.

Other accounts of the time reiterate these observations:

> In 1893, when the Durbar became alarmed at the extinction

> of lions, a rough census was taken, and the number remaining was estimated at twenty-six, which subsequent estimate raised to thirty-one. Since 1893 the number shot is known to be six, including those for which the Durbar gave sanction; and the skins of two others were found in the jungle… There are now estimated to be only twenty lions remaining in the Gir, of which eight are cubs. The extinction of this noble game would seem, therefore, to be only a question of time, and as the Gir is more and more opened up by roads, the rate of extinction will perhaps increase. When less uncommon than it now is, the Indian lion was said to avoid, as a general rule, heavy forest, preferring sandy hills covered with thin scrub and grass, among which it might be stalked on foot without that excessive danger attached to tiger-shooting under similar conditions.

In fact, finding lions to shoot was impossible even for super VIPs. When Prince Albert Victor of Wales came to India in 1890 and arrived in Baroda before going to Kathiawar in search of lions, '[they] proved very scarce.' It was a time when there were endless rumours that African lions were being brought on dhows from Africa to facilitate hunting, especially by princes who took it as a rite of passage. The pressures were enormous and even in the Junagadh hunting grounds, there was a paucity of lions. The prince (also known as the Duke of Clarence) was assured that East African lions would be made available as few wanted him to leave without his bag and two lions were sent by carts in cages for the prince to kill. The lions are said to have come from the Dewan (chief minister) of Junagadh but the duke changed his mind and the poor caged lions were returned. What I wonder is, how often did such events take place in the lion's history and how many of these rumours were actually true?

By 1900, the lion was being talked of in the past tense. 'The Indian lion appears to differ from the African only in the mane of the former being less developed than that of the latter, and in the fact of the black mane, sometimes seen in African lions,

This is another example of a tiger-striped and maned cat; again, the tradition of the tiger was much stronger in everyday life.

never appearing in the case of their Indian cousins. There seems to be no reason for believing that the lions of India and of Africa belong to different species, the slight diversity between them being easily and satisfactorily accounted for by the difference in the nature of their haunts in the two countries. Sterndale gives the length of the lion as 8" [sic] to 9" [sic] feet; and Mr. Selous records the length of two lions shot by him in Africa, and measured between uprights, as 9 feet 11 inches and 9 feet inch respectively. It is sad to reflect that in a few years the lion will be as extinct in India as is the wolf in England, but it is indisputable.'

There were still lions in Persia and Mesopotamia, even though in India it was on the verge of extinction. At this point, the 15-20 lions that were left were intensively managed and reared by the Nawab of Junagadh in his private hunting grounds. As a result, their population slowly rose.

This lion shot on the borders of Bhavnagar and Junagarh was skinned and laid out on a carpet in royal style. This was a shoot meant only for kings and if they were unable to shoot lions in the private hunting grounds of Junagarh, they would shoot them in Africa. The scarcity of the animal meant that credible hunting records were tough to find, and in the nineteenth century, much confusion exists about what was shot and where.

Next page:
As it travelled around the world, the royal crest of the lion found ways to imbibe a variety of cultures while retaining its very royal flavour.

THE AGE OF SLAUGHTER

by VALMIK THAPAR

In the years leading up to independence in 1947, wildlife in the country was mercilessly slaughtered. The main culprits were not just the maharajas, nawabs and princelings of every stripe, but also the British. Some of the accounts of 'hunts' that took place between the late nineteenth century and the mid-twentieth century make for a truly disturbing read. Here, for example, we see a group of valiant royals in Rewah out for a hunt:

> On the way back to Rewah a stop was made at the famous game preserve, a tract of land several square miles in extent, enclosed by a high wall over which not even a tiger could escape. There are three entrances with large gates over which rooms have been built for sportsmen. For a week food in large quantities had been placed in this preserve, and then the gates were closed the day before our arrival. As soon as their Highnesses had taken their places in the rooms, the beat began from the end of the preserve farthest away from the gates. Presently the...victims began to appear, at first in twos and threes and then in large herds. There were hundreds of sambur, but when several dozen had fallen the shooting ceased at the Maharani's request, as she said it was becoming mere butchery. Just then a tigress came along with her cubs. The Maharani had laid aside her rifle, but she

> caught it up again and shot one of the cubs. Instantly the tigress turned and went down the line of beaters striking at them as she went. Five of them were badly mauled and of these three died of their wounds afterwards…

The beaters may have been mauled by the enraged tigress but they were a dispensable commodity; the same was hardly true of the royals, who didn't anticipate the slightest threat from the animals they were destroying wholesale. Indeed, it would be stretching the truth to describe the wildlife parks in which the hunts took place as anything that resembled a real forest. These spaces were to be found all over the country (many of them are now national parks) and were modelled on the age-old concept of hunting parks that had been created for the entertainment of royalty all over the world (including in China, Egypt, Greece, Rome, Mexico), with one significant difference—the weapons of slaughter were now much more advanced so the animals could be dispatched from a safe distance. Everything about the hunt at this time was organized so the royals were not inconvenienced in any way. The royal hunting parks had palaces, pavilions and every luxury possible. As the kings and their queens relaxed, the animals were driven to their weapons. When the Prince of Wales shot in the Terai forests, the Maharaja of Nepal organized hundreds of elephants to create a ring around the game—the circle was tightened before the massacre could begin. These royal hunts only ended in India when our laws changed in the 1960s.

In the 1860s, Louis Rousselet, whose book, *India and its Native Princes*, is a treasure trove of information on Indian royalty, states:

> It is the custom in India for the princes and great nobles to venture on wild beast hunts only when surrounded by so many precautions that they scarcely run a greater risk in killing a tiger than if they were aiming at him from the windows of their palace.

Late in the nineteenth century, the Nizam of Hyderabad, Mahboob Ali Khan, would, during the summer months of April and May, organize hunting expeditions for his guests. The royal saloon was shunted to his private railway station and loaded with food, guns, ammunition and all the paraphernalia for a hunt with goats as bait, and scores of beaters travelling ahead to designated hunting grounds for the kill. The tradition started by the Mughals of stage-managed hunts, where lions were stuffed with opium, was alive and well nearly 400 years later.

This portrait displays the absolute power of the king as he sits on a tiger skin. Hunting lions and tigers were a vital part of the king's life.

After failing to bag any game in the first few shikars, a pair of tigers was bought from a dealer in Bombay, sent out on the same train as the Nizam and tied to trees at the camp. Laudanum was then rubbed into the kill, making them easier to track and shoot.

Other observers of the Nizam's hunt say that they were popular with royalty from around the world.

The dazzling roster of aristocratic visitors to Hyderabad during the early years of the sixth Nizam's rule included the future Czar of Russia, Archdukes of Austria, a Grand Duke of Russia, the Duke of Connaught, and assorted German princelings, all of whom enjoyed hunting in the luxurious game preserves that held every kind of beast including the Nizam's private stock of tigers.

An Englishman who looked after the Nizam's workshop in the 1930s was told by the heir apparent to create a very special machan—one where the royal party was put in a cage from the safety of which tigers could be shot.

> Being a royal cage it had to be spacious enough to accommodate not only the ample person of the Heir Apparent himself, but also several servants, such comforts as ice boxes, alcoholic refreshments, chaises lounges [sic], cushions and, above all, a posse of ladies.

Besides constructing it in sections and reassembling it in the jungle the author adds that a senior member of the Residency staff had suggested that, 'the rigours of camp life are to be tempered with the joys of a portable zenana,' but unfortunately this idea was turned down by his Chief.

Another raja actually designed a vehicle that could be used by the ladies of the court in purdah to take part in the hunt without having to expose themselves to the public.

> Within the vehicle the Raja and his wife would comfortably install themselves for the night of stealthy perambulation up and down the forest tracks, roused from time to time by an electric buzzer which indicated that some roving creature had been caught in the beam of light and stood ready for assassination.

Tigers were naturally one of the most sought-after trophy animals, besides lions, and here the maharaja who outdid everyone else was the Maharaja of Gwalior, Madhav Rao Scindia (1876-1925). He was said to have shot even more tigers than the Maharaja of Surguja whose record was 1,100. Lord Hardinge and his staff shot eight tigers in one beat, Lord Minto shot 24 in three weeks! Every powerful VIP went to Gwalior to hunt and the maharaja 'regarded his tigers as a social and political asset to the state.' This success in Gwalior led to much gossip in those times and many theories were put forth that the tigers were fattened up and displayed a high degree of domesticity implying that they were positioned for the kill either by baiting, doping or driven to the fence of the royal menagerie and 'how it annoyed Scindia!' In their bestselling book, *Lives of the Indian Princes,* Charles Allen and Sharada Dwivedi narrate an incident that took place in Gwalior in the twentieth century. Apparently, 'to ensure docility the tethered buffalos used as live baits for the kills were sometimes sprinkled with opium. Leela Moolgaonkar recalls one such drugged tiger in Gwalior being shot as it snored.'

Just like tigers, I heard recently that, in the past, in Rajasthan, leopards were fed meat laced with opium to the point where they became addicted whereupon they could be shot at given times as they waited for their next dose of the drug!

The Maharaja of Rewah was said to have killed 483 tigers just before and after Independence. His advisor, Conrad Corfield, had his own opinion on the arranged tiger shoot, calling it 'the nadir of shikar because there were occasions when one felt the whole thing was so arranged that there was no sporting element in it.' The absolute power enjoyed by the rulers gave them licence to do anything they wanted, regardless of whether or not it was sporting. Many believed that tigers were caught first and then hunted. Peter Ryhiner, the animal exporter in the 1940s and 1950s, states: 'Tigers are trapped for many reasons: sometimes merely to get rid of a dangerous nuisance, sometimes to provide "bagged" animals for a big tiger shoot, sometimes [in the old days] for arena fights when they were pitted against buffalo, packs of dogs or elephants.'

In a rare book, published in 1920 by Madhav Rao Scindia, the then Maharaja of Gwalior, called *A Guide to Tiger Shooting*, every detail of the stage management of a tiger hunt is not just stated but illustrated to ensure that the tiger has no chance whatsoever of escape. In his introduction, Scindia talks of the confusion caused by the term *'sher'* because it refers to both tigers and lions and then proceeds to talk of the colour differences, one being of a 'bright gold variety' and the other 'of darker hue which verges on the red.' Evidently, there were so few lions about that even their colour needed to be described.

▪

Big cats were not the only form of game to be wantonly massacred—birds, if anything, suffered the most. The Maharaja of Bikaner, Ganga Singh, in 1925 shot 3,300 sandgrouse in one shoot; on another day, in the presence of a viceroy, the count

The slaughter of birds was a frequent event with royalty across the world. Partridges and sandgrouse were a favourite in India.

exceeded 4,300. I have been told that, to set such records, hundreds of beaters would ensure that sandgrouse were not able to find water easily in a huge area; only one watering hole would be kept open for them to drink from and to carry water in their feathers for their young, and when they swarmed to it, they would get massacred. Major-General D. K. Palit describes a hunt that took place in Bikaner in 1958:

> On the day before his shoot he [the manager of the hunt] would send out parties of men with empty kerosene tins, to all the stretches of water for miles around. Their job would be to keep the birds off the water when they tried to approach them in the evening after a day of feeding in the desert. They would keep this up all through the night and into the morning, not allowing the birds to settle on the waters. When HH and his guests had settled into their butts the next morning, the lakes in front of them would be cleared, and the birds would come frantically flocking in to slake their thirst. And the killing would start. There are those

Tiger shooting was pursued with a frenzy in an effort to create vast hunting records and the invention of the car made it even easier.

> who hold that all forms of shooting are immoral but the Bikaner practice seems unsporting to even the most hard-boiled gunman.

Another description of a grouse shoot in Bikaner corroborates Major-General Palit's account, and talks of how the birds were never allowed to land near any tank of water over several days, as orchestrated by the maharaja's men:

> Frenzied with thirst, when the 'shoot' begins they come over in clouds, at desperate speed, to reach the lake... For hours on end the unceasing rattle of guns goes on, suggesting a brisk engagement rather than a sport... Sir Ganga Singh who enjoys it as any guest has been dubbed by a wit king of bikaner...the grouse of god.

Ganga Singh held the record for shooting 25,000 grouse and about the same amount of ducks. The granddaughter of the Maharajah of Kapurthala recalls the frantic screeching of the birds with a shudder: '...it was awful, the poor creatures went berserk

The outcome of a few weeks' shooting could be like a massacre and such slaughters were carefully stage–managed. The British probably exploited India's forests like no one ever had in the history of the region.

because the water was shut off several days beforehand so the air was black with them.'

According to Sir Kenneth Fitz, in Gujarat, blackbucks were polished off by chasing them until they dropped. For shooting duck, the Maharaja of Kashmir '...completed the construction of a veritable floating palace, surmounted by an ornate combination of clock-tower and dove-cot, where he dispensed princely hospitality during the spring and autumn seasons.' Highly trained boatmen drove game to and fro around this floating palace. The maharaja 'would have the first shot and continue throughout the day, a hydroplane kept on the lake being driven around occasionally to keep the duck off the water.'

In 1925, the Maharaja of Tikari invited the Governor of Bihar, Sir Henry Wheeler, for a partridge shoot. For months, birds were bought from the bird markets of Calcutta and Lucknow in huge crates that filled the platform of the station for a month.

> There must have been at least 5000 of them...then a whole area of forest was set aside, ash baths were laid out, corn

> fields were sowed, so that when the partridges were let go they stayed there.

Sir Henry Wheeler shot 500 partridges in four hours and considered it 'the finest partridge shoot he'd had anywhere.'

▪

Unlike the unfortunate game birds which were to be found in large quantities, lions, as we have seen in the previous chapter, were almost extinct, so every royal who could was busy trying to restock his forests with African lions. Lord Curzon, while visiting Gwalior to shoot tigers in 1904, encouraged the maharaja to rear African lions in his territory. The maharaja apparently took a personal interest in the project and allotted an annual budget of Rs 1.5 lakh to it, sending officials to Egypt and Ethiopia bearing Curzon's letter of introduction to try and acquire lions. In time, ten lion cubs were procured and shipped to Bombay.

> Unfortunately two of the cubs died on the voyage and one after arrival in Bombay... [The remaining seven] were reared in Tapovan and next year two of the females gave birth to five cubs. Impressed by the progress four pairs were released in Mohna forest between Gwalior and Shivpuri.

Soon, however, the lions grew tired of wild animals and began attacking livestock. A man was killed by 1910 and in another couple of years, the lions had killed eight to ten people.

> Finally they were caught to be let loose once again in 1915 in a secluded forest area at Sheopur. For some time they flourished there but once again the same story was repeated. They turned man-eaters and had to be shot near their haunts at Neemuc, Panna, Jhansi, Lalitpur and Muraina.

Field Marshal Lord Birdwood had a bizarre story to tell about the Maharaja of Gwalior's repeated attempts to introduce African lions into the country:

> The Maharaja used to tell an amusing shikar story. Gwalior is a large State, with an area of more than 26,000 square

> miles and a population of some 32,000,000, and in the vast forests there are numbers of tigers. Scindia, however, wished to have lion shooting also (the only lions left in India being those in the State of Junagadh in Kathiawar), so he asked Lord Kitchener if he could get him half a dozen young lions from Africa. This was done, and the lions were kept for some months in a very large walled enclosure before being turned out into the jungle to fend for themselves. Unfortunately, they apparently found it difficult to do this, for they took to killing and eating old people coming out of an early morning from the jungle villages. This kind of thing could not, of course, be allowed, and, since the Maharaja was loathe to have his lions shot, it was difficult to know what to do. At all events, the task of capturing the lions was entrusted to a clever Frenchman, by name Auret, who was in charge of the State's shikar department. Auret discovered that the lions had taken up their abode in a small valley, where some old ruins were standing in the jungle, and his first step was to have some huge, strong nets made. Then, so the story goes, he proceeded to lay down an enormous number of powerful and extra-sticky fly-papers in the neighbourhood frequented by the lions. The lions got these on their feet, tried to scrape them off with their teeth, plastered them all over their faces, and thus blinded themselves!—whereupon Auret and his men came along with their nets and rolled up each lion in turn. A very tall story, we all thought; yet the local people declared that it was true. I suggested to Scindia that he might make money out of it, by offering the manufacturers a testimonial as to the lion-holding properties of their fly-papers…we were offered a lion shoot, but, knowing that they were much reduced in numbers, Kitchener declined lest a legend might spring up that he had been responsible for their extermination!

Scindia's obsession with introducing lions to the forests around Gwalior was well known at the time, even though it was obvious that the jungles of Central India did not resemble the habitat of

African lions in any way. I believe that, during the Mughal era, large populations of captive lions were probably bred between Agra and Gwalior and in the private hunting grounds of the emperors. According to Colonel Kesri's memoirs, in the twentieth century, the (then) Maharaja of Gwalior decided to reinstate lions in his territory by importing three pairs of African lions, like many others had probably done before him. They were not interested in Junagadh's leftovers but wanted the black maned lions from Africa. More African lions were imported in the 1920s. The area chosen as their habitat was a 1,000 square mile stretch of land that lay between Sheopur and Shivpuri which was 'divided internally and surrounded by a stone wall twenty feet high broken by strong gates, through which, at intervals, buffaloes were driven. The object of feeding the lions on live meat in this way was to prevent them losing the ability to kill their own food.' The forest that surrounded the enclosure abounded in tigers that seemed to be drawn to the roars of the lions on the other side of the walls; they were often found 'prowling near—probably anxious to challenge the unfamiliar voice. After about four years it was decided to start releasing the lions. They were not all to be let out at once, but a pair at a time.' The first couple disappeared into the wilderness but the second returned to the enclosure soon after they were let go, causing panic amongst the shikaris who brought food for the lions within, as the animals remained near the path where the buffaloes would be driven, ready to attack one as soon as they got the opportunity and begin feeding on the carcass. It was a miracle that they did not attack the shikaris themselves. The next morning, the shikaris and beaters went to the enclosure and made a lot of noise in the hope of driving both lions out of their cover and to discourage their return 'Shortly afterwards however, some of the shikaris…found a trail that led them to the body of the male lion. He had been severely mauled, and since we knew that the first pair of lions had left the area it was morally certain his death was the work of a tiger.' It was clear what had happened: the lions had encroached upon the hunting ground of a tiger, and while the lioness had fled, a fight had ensued between the male lion and the tiger. 'Owing to breeding there were eight

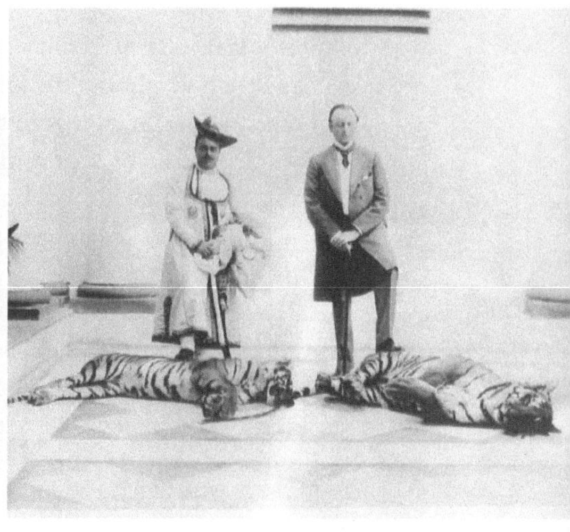

Left: *The shooting of lions for a young prince was a vital rite of passage.* Right: *Gwalior, in the early-twentieth century, is said to have engaged foreign experts to manage the menageries and introduce exotic animals like lions.*

more lions in the enclosure. Six of them were released, two at a time, without incident. Then finally, after a further two months' wait, the turn of the last pair came. These were the only ones that gave any real trouble, but they soon made up for all the rest.' I sometimes wonder how many lions were released in this manner all over India.

A more contemporary observer, Colonel Kesri Singh also believed lions and tigers could not coexist.

▪

In the early years of the twentieth century, despite the attempts of maharajas like Scindia of Gwalior, the only royal who could truly boast possessing lions was the Nawab of Junagadh, in whose private territory the Gir forest fell. As a result, the nawab found that he was in great demand, as there were many VIPs who wanted to shoot a lion. One such sportsman wrote of a lion hunt he was taken on:

> On the 20th (April 1912) we got khabar of a lion, a well-known, unusually big, old and grey lion and started very early after him on foot. I think we were some ten hours

> following him. He at last lay up in a clump of bushes in a very small wood, almost the only wood I saw in Gir Forest. Gir Forest, like a Scotch forest, is treeless and equally bears its name on the lucus a non lucendo principle. I was posted on rising ground behind a bulky tree-stump and the head shikari, a splendid specimen of the class, stood beside me. The 'move' succeeded, but the lion did not wait for the second yell. Immediately [as] he heard the first he bounded out and came past me at a hand-canter. He was inclining towards Robertson's post, when that kind, unselfish young man fired two shots over him, pushing him towards me. I hit the lion rather too high and too much behind to kill him straight away. He went a short distance and then lay up on a bare spot exactly in the attitude of the Nelson Monument lions, and we, the head shikari and myself, had to walk up to and finish him. I fully expected him to charge us and confess to feeling at the moment in a 'blue funk', but my shot had injured the spine and he could not rise, much less charge. He was a grand beast, and when I looked at him lying dead, I confess that I would have given all I have in the world to have been able to restore him to life.
>
> He looked superb when he passed me, taking great leaps rather than galloping; he looked very formidable when sitting up wounded, and he looked noble in death. Of course, I am beyond measure delighted at getting a Gir lion, but the day is not far distant when I shall hate to kill any of God's creatures.

Given his efforts to save the lion, the nawab seemed to feel he had a claim on all lions in the country.

> All the lions were his property irrespective of the area they were found in.

Probably because they had been intensively managed by man for decades, the lions in Gir and surrounding areas behaved like giant dogs, displaying a remarkable familiarity with humans. Colonel Kesri Singh notes:

> I was surprised to notice how little use they made of the available cover. On each occasion the lion strolled out casually into the open, offering an almost unmissable shot. My impression is that these lions are not only astonishingly bold, but, as it were, scornful of taking cover in the face of danger.

Naturally, this kind of behaviour resulted in problems with the locals, especially in Jetpur and parts of the Baroda territory. 'In a short span of 7 months, lions killed 352 domestic animals, 2 women and 1 boy and seriously mauled a man. Lions had become very bold and even entered villages during the day time to attack men and cattle.' There were endless encounters with lions in the area that revealed their dog-like behaviour.

> He came back and bolted the door but that much time was sufficient for a lion, freezing outside, to come into the warm cocoon of the room. After some time, the shining eyes revealed the animal and the shocked and dazed guard woke up his companion… Although, the lion only remained a mute spectator and did not retaliate at all…

In another anecdote, 'the inhabitants…tried their best to drive away the lioness by shouting and throwing stones at her but she refused to budge. She got up later in the afternoon and moved into the empty half of the room.' Such tame behaviour wasn't just witnessed in the villages abutting Gir, there were also cases of lions wandering very close to the city of Junagadh:

> They were audacious enough to wander in the vicinity of the city. Major Carnegy found their tracks on several occasions by the gates of his bungalow, which was in an open area and had a number of other houses nearby. They carried off cattle within a stone's throw of the then new student's quarters of Junagadh college.

There are other examples of their tractability which flies in the face of the behaviour of wild lions elsewhere. C. A. Kincaid was posted in Kathiawar in the judicial service and saw several lions in the Gir hunting preserve of the Nawab of Junagadh. He also got the nawab's permission to shoot two lions and describes the process:

> The Gir foresters' method of marking down a lion is worthy of record. They know, as a rule, all the lions in their part of the forest and their habits. When they are required to produce one they follow it about all night, preventing it from either eating or killing. In the morning it is exhausted and crawls into a thicket to sleep through the day. The foresters then hoist a cot to the top of a neighbouring tree, and half a mile away surround the thicket on all sides except that facing the tree. They send then for the favoured hunter, and when he is seated in his cot drive the lion past him. The object of the cot is not to conceal the hunter, but to give him a good view of the jungle…

Could this have ever been done in Africa?

▪

All this came to an end in the 1960s, when in one fell swoop Prime Minister Indira Gandhi cut the purse strings of Indian royals and rendered them powerless. A ban on hunting followed soon after. However, it was hard for the royal tradition of lion hunting to be totally eradicated. I heard from reliable sources that in 2009 when Bapji, the Maharaja of Jodhpur, went to Africa on safari, it was rumoured that he had shot his first lion there as well. When he returned home, he was greeted in style by 80 of his senior clansmen who came to the airport to greet him in full regalia. They were paying *nazar* (respects) to the lion killer till Bapji told them that he had never shot a lion.

I have been told that, when an earlier maharaja had shot a lion in the 1930s in Africa, there was much celebration and many gifts were offered including the setting up of the Maharaja Kenya Hotel which still exists in Jodhpur. 'It is customary for the head of state to invite the fiance [sic] for a shikar before the wedding. Nothing could be more appropriate than a lion hunt. Only maharajas and the highest government officials ever get an opportunity to shoot the Asiatic lion…many maharajas have

The killing of a lion in Bhavnagar.

never shot a lion.' On 30 December 1940, a lion was 'attracted' into the part of Bhavnagar that borders Gir, perhaps because it was desperately needed for a princely hunt. Interestingly, the hunt succeeded, because even though the lion desperately tried to run back to Gir, a beater on a tree jumped on its back, frightening it towards the hunter—an act that perhaps highlights the tameness and docility of the Indian lion.

Given the enduring connection between royalty and the lion, it was perhaps inevitable that the 'tame lion of India' would meet the fate it has. Prized as a trophy over the centuries, and more so because of its scarcity, the royals could not stop stocking it in menageries and hunting parks, importing it endlessly, and making attempts to breed it. The result was a sort of khichdi lion—a creature that was a curious mix of the North African and the so-called Asiatic lion. Some findings about the mixed up genes of Asiatic lions came to light when the 'pure Asiatic breed' of lions were being carefully bred under a special programme in 38 American zoos and was then genetically analyzed by a specialist team to discover the presence of African forefathers.

We now live in a time of experimentation where captive tigers are being released in places like Madhya Pradesh and have succeeded in propagating themselves. It is possible that similar initiatives must have been tried with the lion as well. In the 1950s, an attempt was made to introduce the Gir hybrid into the Chandraprabha sanctuary near Varanasi in Uttar Pradesh, but the animals were poached shortly after they were released. More contemporary attempts have stalled after the 'Gujaratiness' of the lions became a state issue and Gujarat decided not to share its lions with anyone else.

I personally find this ironic as it is proof—if proof were needed—that the lion's survival in India hinges on its being managed by man. For centuries, it has been bred and hunted by royals in enormous hunting parks that have now been turned into national parks.

The royal family of Gwalior owned what is now Madhav National Park in Shivpuri and vast stretches of forest around it that touched Rajasthan. Similarly, Dachigam National Park (and the chunks of forest that surround it) was once a part of Kashmir; Junagadh had Gir National Park and the adjoining areas; Jaipur had Ranthambhore National Park and the forests that touched Madhya Pradesh; Bharatpur had the Keoladeo National Park; Alwar had the Sariska National Park that nearly stretched till the borders of Jaipur; Kotah (now Kota) had the Darrah National Park; Rewa had the very prestigious Bandhavgarh National Park; Travancore had the Periyar National Park; Mysore had vast areas covered by both Nagarhole and Bandipur; Hyderabad had the Mahavir Harina Vanasthali National Park, Pocharam and the vast Nagarjunasagar Tiger Reserve; Mayurbhanj had, as their private hunting grounds, the enormous Simlipal Tiger Reserve, and the world-renowned Manas National Park served as the hunting park for both Cooch Behar and Gauripur. All these areas probably became greener in the last 50-60 years as the royals could no longer manicure them at their own whims.

Today, there are over 300 lions in Gir after having been carefully managed to bring them back from the brink of extinction, as we have seen earlier, through the efforts of the

Nawab of Junagadh, and later by the Gujarat state government. However, their behaviour over centuries has been decidedly tame, an observation that has also been made by the writer, Kenneth Anderson, who recounts his experience with lions at Gir. He had come across a pride of six lions—two full-grown lionesses and four half-grown cubs—that appeared to be docile and friendly with one another. As the party drew closer to the lions, the lioness, in apparent disgust at their presence, began to drag the carcass of a dead buffalo away. He was amazed at what happened next:

> The two chowkidars ran forward, caught hold of the dead animal by a foreleg, and started to pull in the opposite direction. It was an incredible spectacle. A tug-of-war between two human beings and a wild lioness, with five other lions looking on and a crowd of human spectators. I would never have believed it had anyone told me. The two foresters were no match for the lioness who started dragging them along with the kill. Then, amazingly, they let go of the leg they were pulling and ran forward towards the lioness shouting in unison at the pitch of their lungs. The lioness released her hold, leaped backwards, and stood erect to look at the two men wistfully.

Anderson had reason to be surprised: normally, an alpha predator like the lion in the wild would have made short work of the chowkidars and anybody else in that party who tried to separate them from their prey. The obvious conclusion to draw from this, and other similar incidents, is that neither the Gir lion nor its predecessors is truly 'wild' in nature.

Some of our finest naturalists, such as Kailash Sankhala, the first director of Project Tiger, have in their own way, cast doubts upon the Indian lion being a native species. Writes Sankhala in his book, *The Return of the Tiger*: 'The relationship between the lion and the tiger has been a subject for speculation, and which of the two came to India first is still an open question. One school of thought is...that lions were the original inhabitants and that the tiger entered through the forests of Burma... History does not support this view.' Sankhala goes on to talk about evidence of the

The rite of passage for boys was marked by the killing of wild boars. At the age of eight or nine, the prince's maturity was determined by this process. Hunting wild boars was common even in Assyrian times, 2,500 years ago.

tiger being present in India since about 4500 BCE whereas the lion is of more recent provenance. But he doesn't really come out conclusively in favour of the theory that the Asiatic lion in India has descended from exotic imports from Africa. M. A. Rashid, a senior officer from Gujarat state, is another wildlife expert who theorizes about the origins of the Indian lion, but in my opinion, does not go far enough to conclusively call it an exotic species:

> Evidence of lions in the largely barren interior of Iran, Afghanistan and the westernmost portions of Pakistan is lacking in the past 200 years. Hamun-e-Jaz-Muryan in eastern Iran is regarded as the hottest area in the world for much of the year, while further north, there are daily 185 km per hour winds during part of each year… Lack of records in the intervening range between Iran and Pakistan may be due to the fact that no literate hunters penetrated these regions. However, since lions colonized India from the west, they must have certainly crossed this area… The total absence of lions from the region south of the Narmada further indicates that India was not its original home… This limited distribution of the lion in India also lends support to the theory that it was a more recent migrant into India as compared with the

tiger which is believed to have migrated into India much earlier from its original home in the icy wastes of Siberia and Manchuria through the north-eastern passes.

▪

Before I end this chapter on the lion, I must confess to feeling horrified at the way Indian emperors and other royals have behaved towards our wildlife over centuries. I hadn't realized the extent of their absolute and destructive power until I began researching this book—the endless examples of the slaughter of wildlife in controlled conditions by our royals remains one of the most under-reported facts of Indian natural history. And most importantly, these were mainly pre-planned hunts and did not take place in the deep, dark forests. They were conducted in the lap of luxury and close to settlements or hunting palaces; neither did the royals have to look for animals as the animals were driven to them and sometimes bred and released for the hunt. It was a sport with little risk or danger, carefully planned and executed. Of course, there were exceptions within the community of royals but they were few and far between. Most loved exotics to the extent that some of their palaces (like in Dungarpur in Rajasthan) had wall paintings that depicted numerous African animals, mostly boasting their presence in their private forests.

When the Prince of Wales shot with the King of Nepal, more than 500 elephants squeezed the game into a 'ring' for the massacre. The Nepalese king introduced at least two to four African lions into the terai early in the twentieth century. The Nawab of Junagadh—who not only had diamond collars made for his dogs but invited 50,000 guests when his favourite bitch, Roshana, mated with his golden retriever, Bobby—was the one who bred the Gir population of lions. These were the ones I call a 'mongrel' pack of 10-20 animals that remained at the end of the nineteenth century and were hand-fed and hand-reared in his private hunting ground. The Maharaja of Gwalior imported lions on several occasions from Africa—be it Mombasa, Ethiopia or Mozambique—and other places, with a view to have both tigers and lions in his private forests like few other kings could do. In fact, whether it was Jamnagar, Bharatpur or Alwar, most well-known royal houses had some lions in their menageries. The maharajas loved to experiment. Maharao Khingarji of Kutch and Maharaja Laxman Singh of Dungarpur stocked their private forests in the first part of the twentieth century with leopards and tigers. Dungarpur released nearly 20 tigers into the forests that he controlled for purposes of hunting. I wonder how many lions were shipped into other areas? The maharajas had become proficient at introducing big cats into their private forests. Not to forget, the rhinos that were shipped by sea into Udaipur, red deer into northern India, African buffaloes into Coorg, giraffes and zebras into vast enclosures, right from Gujarat to Lucknow. In fact, between the species and subspecies was a cocktail of genes that would astound anyone (and I haven't included the smaller animals, birds or even exotic plant life). Wild boar were fed and fattened for the kill in Udaipur and Jodhpur and some really huge ones were introduced for the hunt. Absolute power meant the ability to do strange things that were at odds with nature and the royal menageries played a crucial role in this process.

Laura Betzig, a historian, puts it succinctly: 'Kings set aside enormous parks; they stocked them with game; they maintained well caparisoned elephants and horses, and they bred thousands of hawks and dogs.' She goes on to state how 'some kings assembled extravagant menageries with thousands of animals; they hired or maintained tens of thousands of beaters or falconers; they camped in enormous bivouacs, outfitted in silk or sable; they caparisoned the elephants and horses they rode in silver and gold.' During the hunt itself, the kings liked to display their power and status through 'enormous amounts of manpower, horsepower, material resources, and time.'

I believe that the lion was an irresistible and magical creature and an alien imposter in the land of the tiger—a natural world that he didn't know or belong to. Be it to prove their power or to propagate tales of an exotic and royal beast, those who ruled the land kept the myth of the Indian lion alive for over 2,000 years.

THE STORY OF THE CHEETAH

by VALMIK THAPAR

I was lucky to spot seven cheetahs in the wild in the Masai Mara in Kenya and get insights into their natural behaviour. Kenya was one of the greatest suppliers of cheetahs to India. The evidence of this is endless in the twentieth century and therefore must have occurred in earlier times as well when groups of cheetahs were caught and shipped to India

Next page:
The cheetah and his keeper. Most of the cheetahs unaccounted for in India from the sixteenth to the twentieth centuries were on a leash—the royal 'hunting' companion.

The cheetah is commonly known as the Indian hunting leopard—even in Africa, where it is indigenously found—but it is neither a leopard nor is it particularly Indian. In fact, the word 'Indian' stems from the high demand it generated from India. Moreover, although *Acinonyx jubatus* is classified as a member of the cat family, it has many distinctive features that in no way resemble the felines it is grouped with. The professional big game hunter, T. Murray Smith, speaks of the spotted predator in his book, *The Nature of the Beast,* where he marvels at the uniqueness of this 'most strange animal':

> He behaved as though he were still in the Garden of Eden, where there was no hatred and no hunger: 'and to every beast of the earth, and to every fowl of the air, and to everything that creepeth upon the earth wherein there is life…'

Smith further believes that the cheetah was not a true cat and its claws were not fully retractable, but it had the speed of a greyhound. 'He is also dog-like in his docility. I have crept upto a wild cheetah and pulled its tail as it crouched in long grass.'

While I cannot claim to have pulled a cheetah by the tail, I can vouch for the fact that it is a gentle and friendly creature. Once, I observed a mother and her six half-grown cubs in Africa for several days; one day, I was startled when one of the cubs jumped into the empty seat next to mine in the open Land Cruiser from which I was watching the family. It softly snarled at me from inches away, and I fell under the spell of this magnificent animal.

Others are no less susceptible to its speed, grace and tractability. French anthropologist, Odette du Puigaudeau, remarks on the cheetah's character 'which greatly facilitates his training; the royal conviction that wherever he may be he is at home, wherever he may go he is king, a most gracious and detached sovereign whom nobody would dare to attack.' She describes it as 'a companion who combines the calm nobility of the lion, the grace of the panther and goodness of the dog...' Further, she explains how the animal is 'universally known and nowhere feared' and forms the link between the dog and the cat. 'The Arabs assert that the first two cheetahs were born from the union of a dog and a panther.'

As we have seen early on in the book, Mughal Emperor Akbar believed the cheetah was 'one of God's wonders'; it has been a royal pet, and an animal that royalty has hunted with in this country until as recently as the middle of the twentieth century. The pharaohs believed that 'the cheetah, as the fastest animal on land, would carry their spirits away from death.' The Hittites mounted their principal goddess on this animal, it was the preferred motif on Tutankhamen's tomb and in the Pharaonic era, cheetahs with human heads were shown trampling down evil. The Cretans, Greeks and Etruscans used the cheetah for hunting and as a royal pet. Much later, cheetahs were present in royal menageries in Greece, France and Italy and they were used for sport in the fifth century in Italy and then later in Russia, Syria, Palestine, Ethiopia and China (Kublai Khan kept hundreds of cheetahs in his hunting parks). Training cheetahs became somewhat of a trend in Europe between the seventh and the seventeenth century.

■

For the longest time, there were thought to have been six subspecies of the cheetah—five in Africa and one in Asia. The

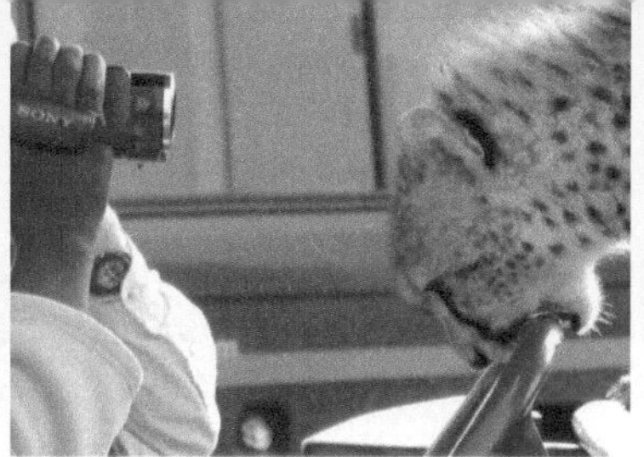

A sub-adult cheetah had climbed into the seat next to me in our jeep in Africa. It sat perfectly still as it stared at me from less than two inches away.

Asiatic cheetah is supposed to have been indigenous to Persia (modern-day Iran, where several dozen are still thought to exist in the wild), India and other parts of Asia, but I believe it is facile to classify the Indian (or Asiatic) cheetah in that way, for it would then mean that the subspecies was indigenous to this part of the world, which, as my co-authors and I have sought to prove throughout this book, is not a certainty by any means. Even the geneticists cannot seem to agree. When India was thinking of reintroducing the cheetah to the wild a few years ago, a conference was held in 2009 to determine whether the Indian and the African cheetah were identical, genetically speaking, which would have made it easier to import African cheetahs and resettle them in the wild. The geneticist, Stephen O'Brien (whom we encountered earlier in the book), while trying to determine the genetic make-up of cheetahs, felt the Indian and the African cheetah had very few differences in their genetic make-up. His findings were disputed by a group of scientists in 2011; in an article published in the journal, *Molecular Ecology*, they claimed that the Indian cheetah was a differentiated subspecies from the one found in Africa, which had separated from its African cousin 30,000 to 70,000 years ago. However, until conclusive evidence is offered by scientists, we have no alternative but to rely on the historical record for our deductions. Cheetahs across the world have so little genetic variation that the analysis of its subspecies is an issue that remains unresolved.

As far as I am concerned, cheetahs in India came into this country as gifts or tributes and were imported by land and sea from Africa and Persia. According to some scientists, a 'bottleneck' in its evolution meant that only a particular type of species was propagated throughout the planet. This is the view I subscribe to. The genetics of the cheetah and all possible subspecies the animal might be divided into is a complex field to explore but judging by what history tells us, the cheetah in India has always been an exotic alien. Since they never bred in captivity, they must have been imported quite frequently and offloaded at ports like Surat in order to keep the royal menageries stocked with them. Some of these cheetahs must have escaped and turned feral while some survived, creating a small 'wild' population that gave rise to the notion that the cheetah existed as a native species in the subcontinent.

From the sixteenth century onwards, I believe most of the 500 or so Indian royals had menageries which contained cheetahs, implying that there must have been 2,000-4,000 cheetahs all over the country in captivity. In the wild they were practically invisible.

Divyabhanusinh has stated that from 1772 to 1967, a total of 230 cheetahs were seen or shot. This averages to about one a year—an astoundingly low figure when compared to the thousands of tame cheetahs that were being herded around on leashes. It was during the time of the Mughals that the mania for the cheetah really caught on. Although cheetahs are not talked about in the *Baburnama* (beyond a single mention of a cheetah keeper), in the time of Akbar and those succeeding him, foreign travellers to India encountered tame cheetahs regularly.

Father Monserrate never saw any wild cheetahs while he travelled to the court of Akbar between 1580 and 1582. However, he saw Akbar hunting with his tame ones, observing how the '...panthers are drawn by horses, under the care of keepers to the place where game is feeding. They are blindfolded so they may not attack anyone on the way. When they are freed, they dash ravenously upon the quarry; for they are kept in a state of starvation.'

The *Ain-i-Akbari* is very clear that one form of hunting was to gallop after these unleashed cheetahs on horseback to chase down deer and wild boar. Many scholars used these paintings in their manuscripts to postulate that depictions of unleashed cheetahs were evidence of large cheetah populations in the wild but this is incorrect.

The Indian wilds were not suitable for the cheetahs. Although Indian royals used them to hunt down blackbucks, there are few accounts of them dispatching chinkara or the four-horned antelope. Besides the absence of suitable prey that is normally found in the grasslands and habitats they are indigenous to, they would have had to contend with the menace of wolves. The density of wolves was so high they would have devoured the cheetahs if they had been around. According to Mahesh Rangarajan, a staggering 2,00,000 wolves were killed between 1875 to 1925 along with 80,000 tigers and 1,50,000 leopards. In the same period, however, only about 40 cheetahs were seen or shot.

In a previous chapter, the much bandied-about notion that Akbar had 1,000 or 9,000 cheetahs has been thoroughly discussed. Whatever the number he did actually have, many must have escaped, either at the time of being trained or on the hunt. Trained animals were very precious and Akbar had special relationships with many of them. Because they were able to get away frequently, special bands of cheetah catchers roamed the countryside looking for escapees. They were trapped in pits—a scene that has been frequently depicted in paintings by chroniclers of that time. There are many depictions of the emperor's concern for the animal's welfare, especially a lovely tale of Akbar travelling from Agra to Gwalior to extract one from a pit:

> One of the joy-increasing occurrences was H.M. the Shāhinshāh's engaging in the hunting of cītah. The lord of the world, though under various forms he appears to be enjoying himself, is in reality carrying on the worship of God. He both tests men and discovers the secrets of the kingdom. With this view he makes hunting a means of gaining knowledge, and employs himself in real devotion. Among these things he especially inclined to the hunting of cītahs, and he has traps made for catching them. The custom was that when news was brought of a cītah having fallen into a trap...he immediately mounted a swift horse and went off to the spot. By proper methods the cītah was brought out from the hole and made over to the skilful

> in the business. On this occasion news was brought that a powerful cītah had fallen into a hole in the neighbourhood of Gwālyār. On the day of Ormazd the 1st Āar, Divine month, corresponding to Sunday 4 Jamāda-al-ākhir, he mounted his horse and proceeded towards Gwālyār. When he came to the hole he himself bound the cītah and took it out.

The story is probably far-fetched as no one talks of how long it took Akbar to travel from Agra to Gwalior, how many accompanied the emperor, how the cheetah survived for days in the pit and whether it was brought back to Agra after it was released. However, what this story portrays is not so much fact as Akbar's concern for his favourite animal.

Besides such exaggerated tales of the cheetahs, many of the paintings of the period do not show much familiarity with the animal in its natural state. For example, very few depict cheetah families—the only miniature painting of a cheetah family from the Mughal period shows two adults with four tiny cubs. This family scene is seldom seen in the African wild—mothers raise cubs in the absence of any other adult. I doubt that this painting was based on reality. Rather tellingly, in over 500 years, there have hardly been any records describing a cheetah with cubs; in contrast, the cheetah with cubs is a common sight all across Africa.

Seventeenth-century French traveller Jean de Thévenot is very clear that cheetahs were trained in Gujarat and regularly sent to the emperor in Agra by the Governor of Gujarat. I am sure they were unloaded at the harbour, acclimatized and trained there.

▪

The rarity of the cheetah is underlined by the fact that, until the medieval period, cheetah art was totally absent in India. It is only in the twelfth century that we begin to find mention and visual depiction of hunting leopards or cheetahs. Art aside, accounts of the cheetah have been confused for a long period of its existence because it can easily be mistaken for a leopard and vice versa. They don't seem to have figured in prehistoric rock art or on Mohenjo-

A family of cheetahs in a rocky landscape—probably the only depiction ever made of a family of cheetahs. Even then it was inaccurate as only the female cheetah is ever found with young cubs (1570).

daro seals, and in the centuries that followed, wild cheetahs find no place in Indian artistic tradition unlike leopards and tigers.

In the 1886 book, *The Indian Empire: Its People, History, and Products*, the author, Sir William Hunter, affirms:

> The cheetah or hunting leopard (*felis jubata*) must be carefully distinguished from the leopard proper. This animal

> appears to be [a] native only of the Deccan, where it is trained for hunting the antelope.

Here, for example, we see Henry Bevan, an old India hand, confusing the two. He is probably referring to leopards in the extract that follows, as cheetahs were not livestock killers or chital killers and certainly would not have lived in the dense forests of Hunsoor near Mysore:

> Of late the cheetahs had become very numerous about the vicinity of Hunsoor, constantly killing cattle while out grazing. As the thickets were too dense and thick to penetrate, to shoot them, traps were set, baited with Pariah dogs or goats, in which several were caught. I lost dogs at different times, when pursuing hares or other game into their haunts. A cheetah also managed to carry off a very fine spotted pet deer, over a high wall, kept in an enclosed yard close to my friend's house.

In another instance, Bevan refers to cheetahs as leopards, but he could be forgiven in this case, because as we know the cheetah is also called the hunting leopard:

> The hunting of the antelope by leopards trained for the chase, is one of the most ancient field-sports in India; packs of cheetahs, or hunting-leopards, were kept by Hindoo rajahs and Mohammedan nabobs, and the European travellers of the sixteenth and seventeenth centuries represent the hunting parties in which these animals are used as the most wondrous exhibitions of oriental magnificence. [On one such hunt that I witnessed] there were three leopards in the field, each carried on a small platform-cart, drawn by two bullocks, hoodwinked and secured with slip collars. They remained so, till we approached a herd of antelope...
>
> As we moved in the direction of the herd...one of the leopards was slipped from off the cart...the beast jumped on the ground, and stealthily, though rapidly, moved towards the antelope, endeavouring to hide his motions by taking

> advantage of every bush and inequality the place afforded... till within a short distance. When he saw the deer bounding away, the leopard put forth his wonderful powers, and was up with them like lightning, or as an arrow shot from its bow, and by a blow from its fore-paw, laid his victim on the field.

Cheetahs were cherished by royalty in the country for centuries. Sir Thomas Roe, who was the first ambassador to India in the seventeenth century on behalf of King James, describes the animals that were part of procession of the kings:

> [First] five Elephants in armour, with banners supported by those that were in their Seats, capable of a dozen Sitters; they manage them by one Eider sitting near his neck, with an iron instrument a Cubit in length, the Point bended downwards as long as a Finger... After these came a Dozen Leopards on State-Hackeries with their Keepers, who train them up to hunting...

Roe describes the cheetahs (or leopards, as he called them) hunting antelopes: 'Antilopes [sic] are set upon by Leopards on this wise; they carry the Leopards on Hackeries, both for less suspicion, and to give them the advantage of their Spring; which if they lose, they follow not their Prey, being for a surprise; wherefore the Hackeries wheel about at a distance, till they come near enough to apprehend them, they feeding fearless of the Hackeries; then with three or four Leaps after a small Chace [sic], seize them, and easily become their Masters.'

In Mughal times, too, the animal was transported with great care. Many miniature paintings show that as courts moved or the emperor went off to various expeditions, cheetahs were transported within covered palanquins, on camelback or on horseback. At times, even bullock carts were used.

▪

Most British accounts of cheetahs in the nineteenth century are still about tame cheetahs used for hunting. Among the first narratives is the following from Captain Thomas Williamson's book, *Oriental Field Sports*. He writes:

> As in hawking, one bird, wild by nature, is taught to pursue and destroy another, so in hunting recourse is had, by the native princes and others of rank, to the seah-goash, and to the cheetah, for the purpose of killing deer, and other game. The seah-goash, literally implying 'black-ears', is a small species of the lynx; its form is beautiful, and from its body, which is of a fine dappled mouse-gray, it becomes gradually blacker towards the extremities, which terminate in a deep chocolate colour. The tips of the ears are of an exquisite finish, being brought to as fine a point as the best miniature pencil composed of sable hairs. Their shape has something peculiarly graceful, and the expression they give at every turn to a most keen and vigilant eye, adds much to their beauty.
>
> The cheetah is a small kind of leopard, or it may perhaps, with more propriety, be considered as a leopard-cat, as many term the seah-goash the tiger-cat. The cheetah is rather an ugly animal, and in lieu of that quick apprehension and animation characteristic in the seah-goash, seems either to view objects with a vacant stare, or to regard them with the most malignant ferocity. One would conclude from its superior weight and apparently greater power, that the cheetah was of the two far superior. Experience, however, justifies the opinion that the seah-goash is, in its wild state, infinitely more destructive. The Sultaun Tippoo had several cheetahs, but as far as I could learn, not one single seah-goash in his collection. If I am rightly informed, it is very difficult to rear them, and more so to fix them in a domestic state; being apt to disappear after glutting with the blood of their prey during which time it is extremely dangerous to attempt securing them. As to hares and foxes, as also jackals, the cheetah and goash, though the latter is scarcely larger than a full grown tom cat, soon overcome them. Deer are their principal object; but extreme caution is requisite

> in managing matters so as to avoid accidents. These savage animals are carried to the hunting ground in cages, conveyed by carts, and on the game being up, the door is opened, when the cheetah, or seah-goash darts forth at speed after the animal in view. They are so extremely fleet, that if the ground be fair, they rarely fail to overtake within four or five hundred yards, when the seah-goash in particular, springs upon the rump of the deer, occasioning it to look back, or to hold up its head; then with a second bound it seizes on the back of the head at the spot where the vertebras of the neck are inserted, and there fixing its teeth, often strikes the prey senseless. I never was present but at one chase of this kind. Curiosity led me, as it did many others, to see what I had never seen before: but I was not much diverted. A deer was turned loose on the occasion, and a seah-goash sent after it. Two minutes finished the hunt. I was not aware of the propensities of these animals to follow horsemen, or any other moving object, when the game might accidentally escape out of sight or else I certainly should have been more diffident on the occasion, and taken my ideas on the subject from some eye witness...

Other observers such as Daniel Johnson make similar comments about the hunting method of the cheetah and caracal in India. Observations of the cheetah hunting down blackbucks are common throughout the century, but the following account from Sir Samuel White Baker's memoir deserves quoting in full as, to my mind, it is unparalleled in the beauty and precision of its descriptions:

> We reached a position within about 120 yards before the two fools [bucks] observed us. They at once left off fighting, and having regarded us in astonishment for half a second, one dashed off to the left, and the other to the right, across the open plain devoid of bush...or any obstacle to the highest speed.
>
> At that same moment a cheetah that had been held in readiness leapt airily to the ground, and the chase

commenced after the right-hand buck, which had a start of about 110 yards. The keeper simply begged us not to follow until he should give the word.

It was a magnificent sight to see the extraordinary speed of both the pursued and the pursuer. The buck flew like a bird along the level surface, followed by the cheetah, who was laying out at full stretch, with its long, thick tail brandishing in the air. They had run about 200 yards, when the keeper gave the word, and away we went as hard as the horses could go over this first-class ground, where no danger of a fall seemed possible. I never saw anything to equal the speed of the buck and cheetah; we were literally nowhere, although we were going as hard as horse-flesh could carry us, but we had a glorious view.

The cheetah was gaining in the course, literally flying along the ground, while the buck was exerting every muscle for life or death in its last race. Presently, after a course of about a quarter of a mile, the buck doubled like a hare, and the cheetah lost ground as it shot ahead, instead of turning quickly, being only about 30 yards in the rear of the buck. Recovering itself, it turned on extra steam, and the race appeared to recommence with increased speed. The cheetah was determined to win, and at this moment the buck made another double, in the hope of shaking off its terrible pursuer; but this time the cheetah ran cunning, and was aware of the former game; it turned as sharp as the buck; gathering itself together for a final effort, it shot forward like an arrow, picked up the distance that remained between them, and in a cloud of dust for one moment we could distinguish two forms. The next instant the buck was on its back, and the cheetah's fangs were fixed like an iron vice upon its throat.

The course run was about 600 yards, and it was worth a special voyage to India only to see that hunt. The cheetah was panting to an extent that made it difficult to retain its hold.

There were a few drops of blood issuing from a prick through the skin of the right haunch, where the cheetah's

The Kolhapur stables of cheetahs where each animal was African in origin. African trainers were also regularly imported by many.

> nails had inflicted a trifling wound when it delivered the usual telling blow of the fore paw, that felled the buck to the ground when going at full speed; beyond this there was no blood, until the keeper cut the throat in the customary manner, and the cheetah, much exhausted, was led to its cage. This was a very exceptional hunt, and a friend who was present declared he had never seen anything to equal it, although he had been all his life in India....

However, although the British sportsmen did not mind witnessing a cheetah hunt down its prey every now and then, it is clear they preferred to hunt them. Wrote one such 'sportsman':

> Sudugee, 19 November—Nothing has been done today, except taking out the Kolapoor Rajah's hunting-cheetahs. They had three runs each at bucks without killing, which I was not sorry for, as this, like falconing and coursing, is a sport by no means to my taste. All I wished to see was the extraordinary degree of speed exerted by the leopard in chase; and this was displayed today in perfection. The rapidity of his stroke, and the length of his bounds, are almost incredible; giving a rate of going, for a few seconds, too rapid for the eye to follow. In judging of the speed exerted, there is only this to guide you—an antelope, one

> of the swiftest animals in nature, going his best pace, and straining every nerve to escape; and the bounding leopard, flying through the air—with a velocity that gains upon his prey as if it were only going at a gentle canter. The reason of their not killing today, was being slipped at too great a distance, the antelope being very shy. Neither of the leopards could or would keep up his prodigious velocity for more than three hundred yards; and failing to strike the buck within this distance, he became sulky, lay down, and remained growling, till the keeper, coming up with his cart, blindfolded and secured him, after having appeased his wrath with a lump of raw meat. This style of hunting is a beautiful sight to see once; but, in my opinion, is a sport better calculated to please an effeminate rajah than a European sportsman.

▪

Many British observers from the nineteenth century onwards reveal how the cheetah was trained to be comfortable in the presence of human beings. Narrates William Sleeman, the famous scourge of the thuggee cult:

> A little after mid-day the rajah and his brother came to pay us a visit; and about four o'clock I went to return it, accompanied by Lieutenant Thomas. As usual he had a nautch (dance) upon carpets, spread upon the sward under awnings, in front of the pavilion, in which we were received. While the women were dancing and singing, a very fine panther was brought in to be shown to us. He had been caught, full grown, two years before; and in the hands of a skilful man was fit for the chase in six months. It was a very beautiful animal, but for the sake of the sport kept wretchedly thin. He seemed especially indifferent to the crowd and the music…

Sleeman writes in another volume of his memoirs:

> In my morning's ride, the day before I left Gwalior, I saw

> a fine leopard standing by the side of the most frequented road, and staring at everyone who passed. It was held by two men, who sat by and talked to it as if it had been a human being. I thought it was an animal for show, and I was about to give them something, when they told me that they were servants of the Maharajah, and were training the leopard to bear the sight and society of man. 'It had,' they said, 'been caught about three months ago in the jungles, where it could never bear the sight of man, or of any animal that it could not prey upon; and must be kept upon the most frequented road till quite tamed. Leopards taken when very young would,' they said, 'do very well as pets, but never answered for hunting; a good leopard for hunting must, before taken, be allowed to be a season or two providing for himself, and living upon the deer he takes in the jungles and plains.'

Interestingly, the trade in cheetahs had moved beyond the palaces and hunting reserves—by this time, there were cheetahs for sale in

Seen above are two caracals (on the far left and right), a cheetah and falcons—all animals of the chase, much encouraged and enjoyed by the maharajas and their clansmen. This twentieth-century picture must clearly be of an African cheetah.

A unique picture of caracals and a cheetah on a charpoy with their keepers in a small town in North India. The history of the caracal as an endogenous species is worth investigating.

Delhi's Chandni Chowk:

> The trumpeting noises of the elephants with the groaning camels varied occasionally by the roaring of a leopard or a cheetah which animals are led about the streets hooded to sell for the purposes of hunting with the unceasing beat of the tom tom, the shrill pipe and the cracked sound of the viol accompanied by the worse voices of the singers are enough to drive a moderately nervous person to desperation.

It was much the same in the city of Hyderabad where cheetahs cost 100 to 250 rupees and were parked in large numbers on the street:

> In the suburbs of the city of Hyderabad I have frequently seen hunting cheetahs exposed for sale; they were picketed by each leg being fastened to a tent peg firmly driven in the ground, and further secured by a chain round the neck. Even in this position, where they literally could not move a foot, they contrived to be nevertheless in a state of perpetual motion, by swinging their bodies to and fro. Even in the town of Hyderabad, I understand, numbers were constantly to be seen picketed together in this manner for sale.

Later in the century, they were even seen in small towns in

Rajasthan. Marianne North, who travelled through Alwar in 1878, mentions this in her autobiography:

> I saw a whole street full of hunting cheetahs and lynxes... the cheetahs are taken out in carts blindfolded and led out within sight of the deer, when they creep up and spring upon them, holding them till the hunter comes up and kills them. They are so intent on their work that they are easily blindfolded and led away again. All those wild beasts are chained to trestle-beds in front of the houses down the street, their keepers sitting or sleeping behind them, and little children, peacocks, cocks and hens, wandering among them without the slightest fear.

Some of the earliest cheetah encounters in the 'wild' reveal their tameness and reflect their flighty nature.

▪

Towards the middle of the nineteenth century, we begin to find the first accounts of cheetahs in the wild. Some of these were clearly escapees from royal menageries, or those which were released for the hunt. This was especially true of cheetahs spotted in the Mysore area. Indeed, the racecourse around the maharaja's palace in Mysore was much like a Roman arena and many cheetahs were speared here on horseback. The local menageries that released the animal for the hunt must have also been responsible for hunters stumbling upon escaped cheetahs in the Mysore and Bangalore forest. Many serious observers conveniently forget what experts said of the cheetah in this area. Big game hunter George P. Sanderson talks of the period between 1864 and 1877: 'The cheetah or hunting leopard is, as I have already shown in the game list of Mysore, almost unknown in the province. During thirteen years I have only seen two skins, both shot by native shikaris. I have never seen the animal in the wild state myself.'

In fact, Sanderson did manage to shoot two cheetahs at Rhowra during his command. He also encountered a cheetah with her cub on another occasion. His dogs promptly attacked the pair:

> One day, shooting at the base of a rocky cliff, a beater on the summit espied a fine cheetah and its cub, lying on a rock a little below him; before I could get to the spot they were disturbed by the ranging of my dogs, and one of them, a large polygar, laid hold of the cub, and was so well seconded by the three terriers, that after a severe clawing they finally overcame it. They were all more or less punished, one of the terriers dying almost immediately from its wounds. I followed the old cheetah, but did not succeed in getting a shot…I only came up with my dogs as the cub was dying; it was about half-grown, and beautifully marked.

Meanwhile, the only record I found of a cheetah attacking a man on foot was that of Captain Outram, who had once grappled with a cheetah that nearly killed him. Outram was one of the most remarkable characters of nineteenth-century India—an expert soldier and hunter, he was trained in the ways of the jungle by the Bhils whose regiment he headed.

Other encounters elsewhere in the country seem to indicate that there were strays living in the wild—it is likely these had descended from the escapees of previous generations that had turned feral and a few may have even bred.

William Rice, a keen observer of animal behaviour in the nineteenth century, found a cheetah which had been attacked by a tiger in hilly terrain. He observes: 'It struck me as rather odd to find a "hunting cheetah" in the hills, for this animal lives mostly in the plains…' This may have been because the cheetah was an escapee from a menagerie that had strayed into the wilds. 'The tiger [had] no doubt, surprised the "cheetah" asleep, for the marks of the tiger's claws, from which blood still was flowing, were quite plain on the body. This animal (harmless enough except for deer) he must have killed in mere wantonness.'

It is clear from Rice's account that the cheetah could never have coexisted with the tiger in India as each would have required its own ecological niche.

▪

A cheetah and its keeper. It was essential that the keepers strapped the animal well to prevent any avenue of escape.

While British hunters left behind graphic details of the tens of thousands of animals they had 'bagged', there are just a few pertaining to cheetahs in the wild. Gordon Cummings, who was a fine shot, managed to shoot only three cheetahs after thirty years of shikar:

> Sometime after my return to Sirdarpore, a man of the Bheel Corps, who was out with Futtah in search of antelopes, came and informed me that [he and his companions] had seen two tigers in some grass lands a few miles east of the cantonment. I was very incredulous, as the place was not one in which tigers were likely to be, though I knew that the spot was occasionally frequented by panthers. I went out, however, with Blowers, and having taken up positions in trees, sent beaters round to drive towards us. Presently two hunting cheetahs came over the hill, and crossed the river at some distance from Blowers, who fired without effect. They then went off over an open country, cultivated here and there with crops of millet. My men pursued them on foot for several miles, keeping them in view, and eventually turning them back towards the river, where they lay up in some high grain. As the men advanced, several jackals broke away, but just as they reached the end of the field, the

> cheetahs bounded out. I fired, and wounded one: however, they went on, and passing through some tall hemp, swam over a deep pool in the river, and went across the grass lands. Here they were headed off by a party of my men. The wounded beast, being unable to keep up with his companion, halted, and, creeping up to a small tuft of grass, lay down. We went up to him, and as he lay ready for a charge, I fired and turned him over. The cheetah, from his great length of limb, stands very high, and an imperfect view of him in the jungle will often leads to his being mistaken for a tiger.

William Rice writes about encountering five cheetahs in his book, "*Indian Game,": (From Quail to Tiger)*:

> On one occasion, [upon] meeting with a batch of five hunting leopards in a good open plain, our party tried to ride them down and spear them, for we were looking for hog at the time; but they were far too quick for our good horses, cantering along just out of spear-reach quite easily, and even looking round or back. On coming to a large, deep, well-wooded ravine they slunk off down it quite leisurely.

Another British sportsman, Major-General D. J. F. Newall, found six cheetahs while looking for a lion:

> I once speared three cheetahs (the hunting leopard) off his ['Rugby's'] back, one after another. I was out on leave by myself, chiefly after lions. My native shikari (game tracker) came into my tent with an auspicious grin, just as I was finishing my breakfast, and said, 'Sahib, I think I have got the big lion for you at last today; I have marked a big beast under a tree about four miles off; what it is I do not know, as I could not find his tracks on the rocky ground, and would not go near for fear of disturbing him; but I have put men in trees round to watch, and now the sun is hot he is not likely to move.' Of course we were soon on our way. I rode an old shooting pony, but 'Rugby', with a hog-spear, was led in close attendance. The country was for the most part

hilly, with deep ravines and between the little rocky hills or knolls were small patches of cultivation, with every here and there very fine trees. From one of these trees a white rag was waved as we came on, telling us all was right so far. The tree under which the lion was supposed to be lying was soon pointed out to me; it was the largest one near, and stood handsomely in the centre of the little plateau by itself, throwing a shade nearly all over the bit of land round it. Leaving the old shikari on the high ground with my spare guns, spear, and 'Rugby' with my rifle cocked, I rode my pony quietly circling round the tree. I made all the use I could of my eyes, but could only make out that there was something very large of reddish-yellow colour under the tree, probably a lion, but I could make out nothing clear enough to justify my firing. At last I got within about thirty yards' distance, and looking intently, saw, as I thought, a large beast lying at full length fast asleep, offering apparently a most lucky shot, and taking deliberate aim at what I took to be behind his shoulder, I fired. To the shot up sprang six cheetahs, beautiful brutes, growling and rushing over each other, one evidently severely wounded. I was really so taken aback myself, I was stupid for a moment; but before I could determine whether to fire my second barrel or to bolt, the old shikari yelled out, 'Come quickly for your horse, they are cheetahs, you can spear them; we'll kill them all!' In another minute I was mounted on Rugby, spear in hand, charging after the biggest of the cheetahs, which were now bolting in different directions. I caught up the one I was after hand over hand, but suddenly the brute crouched, and faced me fiercely, exactly like a tiger waiting for his spring. I did not half like the look of him; I thought the little horse would not either, and but for the old shikari's confident speech to me at starting, I think I should have sheered off and back for my rifle again. 'Rugby', however, had no hesitation whatever: he carried me fast and fair straight at the dangerous looking brute, just as he would have done up to a hog, and I luckily sent my spear straight through behind

> his shoulder, turning him right over. 'Never mind him, leave him to me; there's another to the right,' again shouted the old shikari, who was scuttling after me as fast as he could on his pony; and almost without stopping I at once turned off 'Rugby' after another, which I also speared after a short ran [sic]; and then in like manner a third.
>
> The one I had wounded in the first instance by my shot under the tree had been finished off by the old shikari with my second gun, so four of the six were brought to bag. It was a capital morning's sport, but a very severe run for dear little 'Rugby', who, however, was none the worse for it.

One British shikari was extraordinarily lucky to see a black cheetah. In their memoir, the Wilkinson brothers record the adventure of Sir W. Turner, who writes:

> I was out walking before breakfast with one of our officers. He had a walking-stick and I a gun, which I had just discharged, when a mouse deer sprang out of some bushes and crossed the road in front of us; my dog Whiskey saw it, and immediately gave tongue in chase, but had hardly run a hundred yards when her cry was changed into a sharp yelp, as if in pain or fright. For a moment we stood still, at a loss to imagine what had occurred. Laying down the gun, and snatching my friend's stick, I rushed into the jungle, and again heard a short, stifled yelp not far from me. On reaching the spot there stood a black cheetah with his paw upon the dog, curling up his lips with deep low growls. I felt that I was no match for him with a stick, but was determined that he should not have the dog without a fight; and leaping over the intervening bushes struck at the cheetah with the stick. With one bound he was off, leaving the dog lying on the ground. I carried the dog to a tank that was near, and washed the wounds. Although it recovered it was long before I could get it to leave my heels and again take to the jungle.

But apart from these and some other scattered accounts, few

British hunters managed to see the cheetah in the wild, let alone shoot one. Interestingly, some of them encountered cheetahs close to human habitation which reinforces the idea that these predators had probably escaped from menageries. Sir Edward Braddon, in his book, *Thirty Years of Shikar,* was called upon twice to deal with cheetahs that had been imprisoned in huts while attempting to prey on goats:

> Among the beasts of the Deoghur jungles an occasional cheetah was to be found. In Oudh and other parts of India this animal is domesticated and kept by sporting rajahs for the purpose of running down antelope; in the Deoghur country they kept themselves by running down the goats and sheep of the people. A curious animal is this hunting cheetah—a cat (i.e. a small and much attenuated leopard) down to its feet, and at those extremities a dog. Twice in the course of my Deoghur career was I summoned forth from my cutcherry to shoot cheetahs. In both instances they had been imprisoned in a hut into which they had made their way after the goats of the hut-holder, and as to both I pursued the same tactics—that is to say, I rode gun in hand to the scene of action, from five to ten miles distant, climbed on to the thatched roof that covered the cheetah, and made a hole in the thatch in view to shooting the spotted thief where it crouched below. In both instances I failed of this purpose in consequence of the cheetah's anticipation of my plans; for so soon as I had displaced enough of the roof to make a hole through which I could look into the interior, the cheetah came out by it, and springing to the ground went off. On the second occasion, when, forewarned by previous experience, I conducted my house-breaking with a more jealous care as to monopoly of my skylight, the cheetah was still too many for me, and, bounding out from ...a beam upon which it lay, swept me before it nearly off my coign of vantage. The first cheetah I killed within a hundred yards of the hut; the second was less summarily disposed of...

Mervyn Smith, a British shikari, also writes about the boldness of cheetahs around humans, which can most certainly be attributed to having escaped from captivity:

> At Jeraikela, on the Bengal-Nagpur Railway…one warm moonlight night he drew his charpoy (bedstead) as usual across the entrance of his hut and slept on it. While he was asleep, a hunting leopard crept under his charpoy, seized and killed the deer, and crept back the way it came, drawing the deer after it, and made off to the woods.

■

In the early twentieth century, there was no appreciable change in travellers' accounts of the cheetah, whether domesticated or wild. In his travels in 1900, Pierre Loti encountered tame cheetahs in Udaipur:

> Servants lead tame cheetahs belonging to the king through the streets. They are led on slips so they may become accustomed to crowds. They wear little embroidered caps tied under their chins with a bow…

He goes on:

> One of the tame cheetahs who is on her way back to the palace to sleep, has seated herself on her haunches near the corner of a street.

As we have seen, tame cheetahs could be found in virtually every royal menagerie, whether as small as Deegh or as big as Udaipur, and so it was hardly surprising that many escaped only to be killed later in the wild.

In 1904, sportsman Mervyn Smith killed a cheetah in the dense forests of Chota Nagpur:

> It is generally believed that the cheetah is only found in the more open parts of the scrub jungle of Central India, but I have killed them in the dense forest of Chota Nagpur. The skin is differently marked to that of the panther. Both have a

The royal passenger on a cart, ready for action.

> yellowish brown ground with black spots. The spots on the panther are rosettes: on the cheetah they are skimpy black dabs without a central opening of yellow.

Smith's observations about the cheetah being found in a more 'open' terrain points to the strong possibility that the cheetah that he killed in the dense forest had, in fact, strayed into it, or was an escapee from a hunting park or menagerie.

Notably, both the cheetah and caracal continued to be extremely rare in the wild, as R. G. Burton confirms in 1936. This is what first began to rouse suspicions that the Indian cheetah might not, in fact, be indigenous. *The Imperial Gazetteer of India: The Indian Empire* (first published in 1881) states quite clearly of the cheetah:

> The Hunting Leopard, generally known in Europe as the Cheetah [a name signifying 'spotted' and quite often applied in India to the panther], is placed in a different genus [*cynaelurus*] on account of its claws being only partially

> retractile and of its lighter build. It is not a common animal in India, and would attract little attention but for the circumstance that it has from time immemorial been tamed and used for hunting antelope.

It is clear that coursing with cheetahs was a fashion from very early in the nineteenth century. In his book, *The Wild Sports of India*, Major Henry Shakespear recounts how cheetahs were used and what price they were available at. Later in this chapter, we will see how they were sourced from Africa to service the needs of the rich; because of the numbers imported, many must have escaped and had to be re-caught or shot accidently. In the *End of a Trail*, Divyabhanusinh records a cheetah being spotted at the Adikmet grasslands near Osmania University in Hyderabad as it was chasing a hare. Possibly an old animal that was released after it had finished serving its master, it 'was sighted again after several weeks, at a nearby village. Here it stayed on and enjoyed the hospitality of the villagers, who either out of pity or for fun fed it on scraps of meat from the local butcher.' However, when someone sighted the animal and informed the king, he ordered it to be caged and fed it as much meat as it could eat. 'Some people say that this animal brought a plea before the king that since he [the cheetah] had provided good sport and fun for one and all, now that he could hunt no more and was famished, it was but proper that Man should take care of him…'

▪

So what was the source of the thousands of India's cheetahs over the centuries? As far as my research has shown, it had to be Africa for the most part. According to Divyabhanusinh, both Kolhapur and Bhavnagar imported 100 cheetahs from Africa between 1918 and 1945. He adds that another 100 were imported by other princely houses and suggests 200 were imported in 27 years.

A big game hunter, J. C. Gandar Dower, was known to have recorded the fact that cheetahs were being raised on a farm in Kenya for export to India and England where cheetah racing was a brief attraction for noblemen.

> They were being bred and trained for export to India, where blackbuck hunting by cheetahs has been a recognized sport for many centuries. The Indian cheetah was in danger of becoming extinct, so an Indian prince decided to replenish his cheetah stable from a farm in Kenya. When Mr. Gandar Dower saw the animals in action he accepted the suggestion that a team of twelve should be sent to England and tested out on some of our large greyhound tracks. At the Romford Stadium the cheetahs took part in three tests…

My research shows that at least 1,200-1,500 cheetahs could have been imported into India in the twentieth century to facilitate royal hunts. Most scholars and authors have almost reluctantly admitted to cheetah imports from other lands. Divyabhanusinh writes:

> …the Baroda manual [on cheetahs] goes on to say that in Africa they did not capture a big cheetah. Instead, they shot it with a gun and then chased the young ones on horseback or with dogs, tired them out and then caught them and sent them to India for training and coursing.

He goes on to state:

> Further, if an animal from Africa had to be purchased it needed to be bought with its cage as it was not likely to get used to a new one soon. This apparently was the only concession made for the cheetahs from across the Arabian Sea in the matter of preparing them for coursing!

On the cheetahs owned by Tipu Sultan, Divyabhanusinh feels that there is a strong possibility of them having been imported as Madras was an important centre of commerce and Tipu Sultan had a flair for the uncommon. He postulates:

> The possibility of these animals having come from Africa cannot be ruled out altogether, though it is unlikely.

Almost as an afterthought, he adds:

> It can be argued that some trained cheetahs along with caracals, lynxes, falcons and hawks may definitely have

reached the menageries of the Muslim invaders from Central Asia or Persia.

Other wildlife experts have expressed serious doubts about the cheetah being indigenous to India. Kailash Sankhala and V. D. Sharma, both of who became the chief wildlife wardens of Rajasthan—cheetah country, if ever there was one—said the same in a paper delivered to the Cat Specialist Group in 1984:

> Existence of the cheetah in the wild in Rajasthan at any stage is doubtful. During the grand period of shikar of Maharaja Ram Singh, 1835-1880...when human population was low, extensive wastelands were teeming with blackbucks and chinkara. The autocratic rule of the Maharaja gave rigid protection to all forms of wildlife. There could be no better period for the existence of the hunting cheetah in wild.
>
> Jaipur, during early nineteenth century, was an important training centre for hunting cheetah (Sterndale, 1884) and a section of the city of Jaipur is still known as Cheetawala ka Mohalla. The cheetah trainers, Chhuttan Khan and Chhote Khan, both centurions...vividly remember that their wards used to be imported from Kabul, Afghanistan and Africa. They inform that there were no wild cheetahs in the princely State of Jaipur. The Cheetah School of Jaipur continued to get support for keeping the art alive, till 1922. It trained cheetah[s] for princes of Saurashtra and of other Indian states.
>
> According to our correspondence with Maharaja of Kolhapur, a state known for cheetah hunting during the princely order, the cheetahs were always imported from East Africa. Baroda was another state where the sport of running down black bucks had achieved perfection, but the State's hunting cheetahs were always brought from Mombasa through a dealer.
>
> Among the princely states of Saurashtra (now Gujarat), Bhavanagar and Wankaner were equally famous for organizing the hunting of blackbucks by using cheetahs. Bhavanagar had its agent in Mombasa, Kenya, who regularly

This blackbuck carriage waits for its cheetah!

> supplied the menageries with African cheetahs. The same is true of Wankaner. The entire northwestern plain of India—Haryana, Rajasthan and Gujarat—were not only similar but better suited in terms of habitat than the present East African plains where cheetah still exists. But no cheetah was found in these Indian plains. I would like to continue the argument but in the present context, it would be sufficient to conclude here that the hunting cheetah never existed in the wild in Rajasthan.

I believe Sankhala and Sharma were quite correct and, sadly, their research carried out more than 25 years ago has been largely ignored. Rajasthan and Gujarat, where cheetahs were said to have existed in the wild, was dominated by packs of wolves which makes it impossible for the spotted cat to have existed there:

> Hopes for reintroduction of the hunting cheetah are not encouraging since its existence in the past in these localities is doubtful.

▪

At the turn of the twentieth century, author and naturalist, Frank Finn, had noted that cheetahs used for coursing were being sourced from all over Africa; so had Pocock, a sportsman, who declares that specimens for the sport were regularly imported from Africa. Furthermore, in his book, *Records of Big Game*, Ward mentions a certain Raymond Hook of Nanyuki, Kenya, who 'made a living by catching cheetahs in east Africa for Arab sheikhs and Indian princes. These animals were used all over India for coursing.' Hook sold cheetahs to the Maharaja of Kolhapur, for example, for 25-30 pounds a head.

Two Africans in Akbar's court with a cheetah as a tribute (1570). Was this the African connection to the cheetah imports?

As it was not possible to breed cheetahs in captivity, it was necessary for Hook to catch them in the wild and then tame and train them in captivity before selling them in India. To catch them, he hunted them to exhaustion with dogs, which then surrounded them in a circle so they could not escape. What happened next was witnessed on one occasion by John Pinfold, and makes for a thrilling read:

> Hook...divested himself of his belt and stirrup leathers. Armed with these he made a pass like a matador at the creature's head. It growled and spat in sullen fury. But the feint succeeded. For a second its attention was withdrawn from Ali [Hook's African assistant], and this was his chance. As he grabbed the tail, Hook dived headlong for the cheetah's neck.
>
> The next few moments will live in my memory. Having grabbed its neck Hook calmly permitted the cheetah to chew his arm, while he muzzled its jaws with the belt. Two more boys then ran in to assist, and together they succeeded in trussing the writhing cat and tying it to a tree.

When Hook finally rose to his feet he was covered in sweat and blood.

'Only a scratch,' he panted, 'can't expect to muzzle a wild cheetah without a sustaining a bit of damage. And the cheetah was worth it. I shall enjoy training her. She shall go to India.'

Most of the training was done in Kenya before the animals were shipped to India. According to the Belgian naturalist, Armand Denis, 'the cheetah is really rather a stupid animal' who tames 'fairly easily', but it could nevertheless be a lengthy process, taking as long as six months. They had to trust their handler completely and Hook had great patience with them. A lot of the training was done with food, and initially the cheetahs were kept hooded until their trainer approached them with food so that they grew to associate him with feeding. Once they were tame they had collars put on and when they had got used to that then a lead was put on and they would be taught to go for walks just as with a dog. They were then trained to go in boxes and go hunting. A piece of rabbit or hare was attached to a rope and it was dragged along the ground and the cheetah would chase it and eat the piece of meat as a reward. Once it got the idea the rope was given to a man riding a horse who gauged the speed of the cheetah so that it didn't give up, and the distance was slowly increased until it was running a fair distance. The cheetahs were then ready for hunting buck. When they saw the buck they were let out and they chased it. If they didn't catch it within one thousand yards they gave up and lay down and waited for the cart with the box on top to collect them; they were then given food and then taken home.

Once the cheetahs were trained, they were sold to an Indian trader in Nanyuki named Osman Allu, who 'paid on the nail'. Behind his shop lived a 'mild little Indian' who made occasional trips to Bombay, taking crate loads of cheetahs with him and 'cooing to them and calling them his beautiful children', so that by the time they reached India they would be obedient to any master.

> In the early stages of the Second World War Hook served as an intelligence officer with the British forces in Somalia and Abyssinia, but returned home to his farm after contracting some kind of obscure disease. In the 1950s he was still catching cheetahs and selling them to those few Indian princes who were still employing them for hunting.

Raymond Hook was quite a character and had all sorts of theories about wild animals, especially the cheetah, which was one of his favourites. His long-standing association with the animal began when he began trapping them for his own amusement and later, selling them to the rajas for their hunting pleasures—his trained cheetahs soon became indispensable in a royal hunt and his popularity grew. 'Meantime his study of the cheetah in ancient times had caused him to make a strange discovery. The beautiful spotted hunting leopard, which over a short distance is the fastest beast on earth, appeared frequently on the monuments whenever civilization had reached some sort of peak. As soon as the civilization declined it swiftly disappeared.'

What is remarkable is that the animal was trained in Africa and not in India. Hook's method of training of adult wild cheetahs has been echoed by other observers, for apparently it was impossible to train cheetah cubs to hunt. Wrote Mervyn Smith: 'For purposes of hunting the antelope and other small deer, the cheetah must be caught when full grown and then domesticated. When taken as cubs they never learn to hunt.'

In his book, *My Sporting Memories,* naturalist Nigel Woodyatt reiterates the need for hunting cheetahs to be pressed into service for the hunt only when fully grown:

> It is commonly supposed that the cheetah is trained to pull down buck by the natives. As a matter of fact it is only tamed by them. A baby cheetah would be useless, as it had never learnt to kill in its natural state, and you could not well teach it how to seize its prey, and also handle it at the same time!... The cheetah's great asset is its first lightning rush, which it has learnt of course in its wild state. It consists of incredibly fast bounds. The longer the animal is

> in captivity the more these rushes deteriorate in pace...
> The ordinary method of cheetah hunting is to take the animal out in a country bullock cart, leashed and hooded. Black buck being used to such carts, will allow them to get quite close without suspecting any danger. When judged to be near enough to the game, the antelope is shown to the cheetah, and the latter slipped. Even with a first-class animal of recent capture there are a good many failures.

▪

Besides ordering animals like cheetahs and other exotics from Africa, numerous Indian royals travelled to that continent to hunt. John A. Hunter's autobiography, *Hunter,* describes one such royal visitor:

> I was guiding a raja who never moved without his personal secretary and doctor. The doctor carried a regular medical clinic about with him, most of the drugs being potent aphrodisiacs as the raja was afraid of losing his manhood from the hardships of African life...

In another book, Hunter, who led many prominent safaris in Africa, says of Indian royals:

> These fabulously wealthy men have always been encouraged to come to the game fields of Kenya because of their lavish spending power... [Royals] with a special fondness for the safari always seek sport in Africa sooner or later. Naturally with their practically unlimited wealth they are able to indulge whatever fancies they have...

Hunter goes on to talk of a maharaja of a southern state inclined to corpulence and advised to lose weight but couldn't in India as he went 'through the Indian jungle in an armoured car, from which he quickly shot a hundred tigers without any effect at all on his weight...' When the maharaja's equipment began to arrive in Africa, Hunter was stunned: 'Vehicles [including an armoured car] had been shipped from England and America and included a mobile wireless receiving and transmitting station, a van for cinematography…an elaborately equipped kitchen, generating plants, lorries…a medical van with an x-ray unit…' As if this wasn't enough, 'the caravan hauled on an articulated trailer powered by a Scammel engine… There were also a number of Elsan chemical lavatories and gaily striped canvas booths to put them in...'

While the party was out hunting, 'a motor-cyclist dispatch rider was sent back to the base convoy bearing full details of the bag' at the end of the day, which was then transmitted to Nairobi and consequently sent to India via a permanent short-wave transmitter. 'It would then appear as a news item in the papers of the Maharaja's state the following day. "His Highness today shot a buffalo, a kudu and a fine black-maned lion."'

Meanwhile, no detail was spared to ensure the personal comfort of the maharaja: 'His Highness sat on the carved chair that had been brought from his caravan, wearing a gorgeous silk robe, and with the jewel in his turban flashing like fire.' At the end of his safari: 'The convoy had all been loaded on to the ship and he himself went aboard the day before she sailed… He drove to the quay in a magnificent Rolls Royce that had already been furnished with lion skin rugs and upholstery selected from the pelts he had shot himself…'

Turmoil throughout the Indian subcontinent because of the struggle for Independence did not affect the hunting exploits of the maharajas, of which the African safari was a staple. The lion was much easier to hunt there and vital for the king. The royals both hunted and paid for the capture of animals like cheetahs to take back home.

Alistair Scobie, a sportsman and hunter in the African bush, writes about his meeting with a royal:

> The maharaja was a man with one simple idea in his head – he wanted to catch a number of cheetahs and to train them to course. 'In my country the cheetah is tamed and used as a hunting dog' he said…

Scobie talks of big farms that were kept by hunters and game catchers to service international needs. Carr Hartley's farm, for

African cheetahs on Indian charpoys! (Bhavnagar, 1940)

instance, was considered one of the biggest traders in the region.

> Not only big game is in demand—antelopes of all kinds, from the rare bongo to the oryx and Thomsons gazelle, are constantly required. Giraffes are often ordered and the black rhino is a sure sale. Leopards and lions are asked for…

This is one of many sources that show that cheetahs were continuously being imported into India. An article published in the 4 December 1937 edition of *The Argus*, a newspaper in Melbourne, categorically states:

> Royal cheetahs have been used for hunting for hundreds of years in India. The Maharaja of Kolhapur still keeps up his hunting cheetahs. It costs him 10,000 pounds a year because all the cheetahs have to be imported from Africa.

A few decades later, the export of cheetahs was beginning to cause serious concern. In their book, *Saving the Game*, authors

Royals with their prized pet, getting ready for the hunt.

Anthony Cullen and Sydney Downey reveal the plight of the cheetah:

> What follows therefore, is an approximate average, providing an indication of the extent of the export trade...23 cheetah from Tanganyika in 1952 and 16 from Kenya in 1956-57...it is absolutely criminal to export cheetah.

It was not until 1960 that new controls on export came into being and staunched the haemorrhaging of African wildlife. When Prime Minister Indira Gandhi abolished the privileges of princes, their excesses involving animals finally came to an end.

▪

Officially the last three Indian cheetahs were shot in 1948 by the Maharaja of Surguja, Ramanuj Singh. These could have been nothing else but imported African animals that had either been released or had escaped. There was a huge outcry at their death.

The photo of the dead cheetahs was sent to the Bombay Natural History Society whose editors 'were so nauseated by this kind of slaughter that their first impulse was to consign it to the wastepaper basket. That anybody with the slightest claim to sportsmanship—and the general run of Indian princes justly prided themselves on that [—] should be so grossly ignorant of the present state...what adds to the heinousness of the episode is that the slaughter was done by motoring through the forest at night, presumably with the aid of powerful headlights or a

Racing with cheetahs after blackbucks in cars (Bhavnagar, 1940).

spotlight. This, it will be recognized, is not only against all ethics of sport but it is a statutory offence deserving of drastic action by those whose business it should be to enforce the law.'

Although the cheetah was now officially 'extinct' in India, many 'sightings' of it in the wild continued for the next couple of decades, probably because they could still be found in captivity until then; some escaped and the others were released by their owners when they could no longer afford their upkeep. Because they bred with great difficulty, it was impossible to replicate the Gir-like situation and the last rumours of cheetahs, both in the wild and in captivity, faded as stringent laws on the import and export of animals were enforced throughout the world.

Despite the cheetah's long association with India, for all the reasons we've seen—its overwhelming presence in the courts of the royals, its scarcity in the wild, the inhospitality of the terrain and the presence of other more powerful predators and unsuitable prey species, among others—it was never, in my opinion, indigenous to this country. That does not, in any way, lessen my affection for this extraordinary animal. Its gentleness, beauty, power and speed, as well as the unforgettable manner in which it moves in for the kill, captivated me when I encountered it near the Talek River in the Masai Mara. After the time I spent with the cheetahs in Africa, I knew that I had to know more about this remarkable animal. This book started but a few months later.

Anything was possible in private hunting parks, in terms of which animals could be hunted at what time. These were 'paradise parks' and ancestral hunting grounds that were much treasured and fiercely protected. I believe this is where animals like lions were stocked and bred when necessary for the hunt.

EPILOGUE

I am a naturalist, not an academic, so I wasn't setting out to write a scholarly thesis when I began *Exotic Aliens*. What I really wanted to do was investigate a mystery I had stumbled upon as I was researching my forthcoming book, *Tiger Fire*. That conundrum, which I've stated at the outset, was that the number of lions and cheetahs encountered in this country throughout recorded history never quite added up. The more I read and pondered over this, the more I realized that the myths about these animals had been propagated by people of influence to whom the 'Indianness' of these species was important for reasons that we have discussed all through this book.

As I refined my arguments to dispel the myths that surrounded the lion and cheetah, something else occurred to me—there have always been two types of India: the Wild and the Tame. For thousands of years, much of the subcontinent was Wild India with thick and inaccessible forests inhabited by fierce and aggressive tribes and wild animals. Indeed, much of it was unexplored and stories and epics portrayed these little-known parts of the country as being inhabited by ghosts and demons and where people were banished to as punishment. Over time, kings and nobles created parks and hunting grounds on the edges of Wild India or close to their own palaces so they could rear wild animals or hunt them in comfort. These were enormous in size and could exceed 1,000 square kilometres. This was Tame India—a diminutive and poor interpretation of Wild India.

By conservative estimate, Tame India took shape more than 2,000 years ago. The Greek military writer, Aelian (170-230 BCE), related that Indian kings had large royal parks more grandiose than those of Persian kings at Susa and Ecbatana (modern Hamadan). He says: '[These people] were provided with man-made lakes, trees, shrubs, and of course birds, fish and animals many of which were imported from abroad.' The story of the Indian lion and cheetah largely took place in the Tame India of man-managed forests around settlements controlled by local rajas, nawabs and maharajas. Few of these royals entered the realm of Wild India. For them, the forest had to be a luxurious pleasure ground and not a place where leeches, snakes, scorpions, thorny bushes and thick undergrowth could inconvenience you. Ganga Singh, the Maharaja of Bikaner, even created canals and waterbodies for the wildfowl he intended to massacre—the more wild grouse he attracted to his hunting grounds, the more his 'stature' grew. The behaviour of virtually all the royals in the country in this regard was no different. In Tame India, animals could be stocked, restocked, herded, driven, surrounded in specified areas and sometimes even introduced; this was where predators such as tigers and lions were baited, hand-fed and often drugged. Typically, these places were ancestral hunting grounds associated with much ritual and privilege (be it initiation rites for princes where lions were concerned, or the right to shoot tigers). Even though the chroniclers of those times often misinterpreted the facts of the hunt, it is clear from an analysis of paintings of the time that these hunting parks were close to settlements, ditched and fenced where necessary, and filled with hundreds of courtiers and huntsmen to assist with the hunt. In addition to the glorious hunting palaces where loyalty lolled about, stockades, nets, cages and pits were all a part of the landscape. Here, you shot for pleasure, without risk, and if you could bag exotics as prey, your stock went up in the eyes of your peers, guests and rulers. Indeed, kings would go to remarkable lengths to enrich their hunting grounds with the import of exotics like African lions. That is exactly what Gwalior did in his private hunting grounds at Shivpuri, which made him unique as he could boast that he possessed both the lion and the tiger. Moreover, by killing these animals, royals were given power over their subjects who thought this display conferred on rulers the power to rid the land of evil and danger.

In the late-eighteenth century, the British were the first to make serious inroads into Wild India in their search for the timber and natural wealth they needed to build their shipping fleets, the railway and to fuel the industrial revolution. They went to places that few royals or their subjects had ever been. In a way, they stepped on the toes of the royals because as they tried to exploit and develop the countryside and its environs, they often came into conflict with the rajas and maharajas over ancestral hunting grounds. Orchha, for example, had island-like hunting grounds—often the one most preferred by royals as the escape routes of animals could be sealed off—such as Karkigarh, which were zealously protected. Even though it was the last place in the world where you could hope to find tigers, there were accounts of them being shot there—it was rumoured that the area was stocked with tigers before a hunt was scheduled. In the twentieth century, Orchha fought British plans to take over Karkigarh for an irrigation project. Similarly, the Jaipur royals fought to preserve Ranthambore as their private hunting ground until it was declared a sanctuary in the late 1950s—a part of Tame India, it is today, one of our finest tiger reserves.

In addition to their battles with the royals, when the British entered Wild India, they came into conflict with forest tribes whom they classified as savage. Their scuffles with the forest tribes often spilled over into skirmishes between the royals and the tribals, who had largely gone their separate ways before the colonials arrived on the scene. The British also damaged and destroyed India's forests to a greater degree than the royals ever did. And, as people like Sir James Outram (in the nineteenth century) and Jim Corbett (in the twentieth) walked through Wild India, the narratives that featured wildlife began to change. Now there were more surprises to be encountered in tales of the hunt—danger and excitement began to seep into these stories.

Viceroys still mingled with kings in Tame India but now many of their countrymen were exploring Wild India, and as they did, its borders shrank. All through this period of great change, the royals clung on to Tame India. Until the very end—that is, till the 1960s, when bans were imposed on hunting and privy purses vanished—India's royals continued to exercise absolute power in their private forests and parks as they managed birds, introduced tigers, imported African lions, leopards and pretty much anything that caught their fancy. Recently, when the Government of Rajasthan took credit for the successful relocation of tigers from

Left: *The African lion—Black-maned lions were probably always imported for breeding. In Gujarat, it was called 'ootia bagh' or 'camel-coloured tiger', revealing a lack of tradition.*
Right: *An old and hungry Gir lion. Nearly two dozen doctorates have been done around what must surely be the most docile pack of lions in the world and their habitat.*

Ranthambhore to Sariska, the Dungarpur royal family quickly contradicted them claiming that they had stocked their forests in the 1920s with more than 20 tigers from Gwalior!

As a conservationist, I was horrified by the behaviour of both India's royals as well as the British towards India's natural resources and its wildlife in particular. Although honourable exceptions did exist, for the most part, the British and the maharajas massacred vast quantities of wildlife, and often in the least sportsmanlike way possible. There is no way in which the slaughter of animals, torturing them for the amusement of jaded royals and sahibs and the suffering they experienced on long voyages in abysmal conditions could ever be condoned or rationalized. We are still paying the price for the depredations of these indulgent 'sportsmen'.

▪

And so, at the end of our deliberations, there are certain conclusions that can be clearly stated. With regard to the Indian lion, these would be as follows:

1. We have established that, not only was there an ancient trade in wild animals, but lions were transported via the seas in every direction. In addition, lions travelled on land routes from Balkh in Turkmenistan in the sixth century CE to destinations like India.
2. The lion has been present in Indian art, iconography and scriptures because it is a symbol of royalty and divine power; this does not necessarily mean that it was a native species to the country, or that it was common. Lion art was all-pervasive in India, much like it was in China, Europe and many parts of the world where no lions existed. The art spread because of the link between royalty and the lion.
3. Because it was emblematic of royalty, the presence of lions was necessary to help with the initiation of young princes into the roles they were destined for. For this reason, the lion was propagated in royal hunting reserves and was encountered in the courts of nawabs and maharajas. It was also hunted in what I call 'canned hunts'—intensively managed hunts on royal hunting grounds.
4. The lion has always been a rarity in the subcontinent. Hunting records and the accounts of travellers—from pre-Islamic times through Islamic dynastics that preceded the Mughal Empire to the Mughals and the early days of the British in India—never speak of the lion in the way they do of the tiger. Lions are rarely encountered, and even when they are part of the 'game bag', their numbers are minuscule when compared to other wildlife. There are more accounts of lions in captivity than those of lions encountered in the wild between the sixteenth and nineteenth centuries. And often when the lion is mentioned, it is actually the tiger that is being referred to, because in many Indian languages the word for both predators is the same.
5. From the eighteenth century onwards, wild lions seem to have only been found in the vicinity of ancient royal hunting reserves and parks, leading to the conclusion that they were feral descendants of man-managed populations. Just like the Gir population of 300 was propagated from 15 to 20 animals,

there must have been instances of the same happening earlier in history.

6. Indian royals had absolute power to do what they wished and importing and trying to breed exotic animals was the least of it. Lions probably did well in this country because, in some places, local conditions and temperatures were similar to that in Africa.
7. Genetic studies have shown that the so-called Asiatic or Indian lion, suffers from severe inbreeding and could be a 'mongrel' in terms of its African connection.
8. The superficial differences between the Asiatic or Indian lion and the North African lion—such as its colouring or the absence of a mane—are probably an adaptation to the Indian climate, environment and inbreeding; there is little possibility of it qualifying as a distinct subspecies.
9. Its tame, almost domesticated behaviour is startling when contrasted with lions in the wild in Africa, and can only be accounted for by the fact that man has fed and managed lions in India for centuries.
10. It would have been impossible for the Indian lion to survive in much of the Indian jungle habitat that is largely dominated by the tiger, which, being more powerful, agile and a master of the thick forest, would have destroyed the lion. In grasslands and bush country, which would have been their natural habitat should they have been an indigenous species, there is a noted absence of the variety of prey that has allowed the lion to flourish in Africa. Even in the early seventeenth century, there were no lions to be seen in places like Daman and Diu that are just a few dozen kilometres from present-day Gir. It would have been very difficult for the lion to survive in this country on indigenous prey species like blackbuck and cinkara unless livestock regularly augmented its diet.

As to the cheetahs, it is clear to me that:

1. In India's history, if cheetahs were seen in the wild, they were escapees from wealthy homes and royal menageries. Most portraits of the cheetah show them secured by leashes, in royal courts or as a part of royal processions. The paucity of cheetah

The tamed and trained cheetah—a prized possession for royalty and the nobility.

art, compared to the depictions of other birds and animals, is startling.

2. We have seen that huge menageries of cheetahs (like the one Akbar had) would have been impossible to maintain in India as we could never have had a prey base of millions and millions of blackbuck units to support such populations. They had to have been imported like dogs were—as animals of the chase.
3. The portraits of cheetahs are mirrored in written accounts. Hundreds if not thousands of cheetahs were observed being paraded around on leashes or in carts than were ever seen in the wild. Also, just like the lion was mixed up with the tiger, the cheetah was confused with the leopard in various accounts.
4. A few cheetahs that escaped could have formed small feral populations but even this was most unlikely and rare as they bred with difficulty. Just a quick glance at some places where cheetahs were shot—from Chota Nagpur's thick forests (in what is now Bihar) to Hunsur in Mysore—point to the fact that these animals could only have been escapees from royal menageries.

5. India's terrain and environment was highly unsuitable for the cheetah—from the uneven ground of our grasslands that would have been problematic for them to run on, to the hostile environment of wolves, hyenas, tigers and leopards that would have made the life of this fragile predator a living hell had it existed in sufficient numbers as an indigenous species.
6. Cheetahs, therefore, must have been imported into India just like lions were by land and by sea; because cheetahs had difficulty breeding, many more cheetahs were imported as compared to lions. The scarcity of official and cargo records in earlier centuries has been a major hindrance for those looking for proof of their origins, but enough printed accounts exist, as we have seen, that point to a roaring trade in cheetahs.
7. The cheetah was a royal pet and companion, one more exciting than the dog. It was, after all, the fastest animal on land. The sport of coursing with cheetahs was an imported fashion. India did not have a tradition of hunting with cheetahs and this is corroborated by several accounts, especially those detailing how they were trained for hunts.
8. Africa was a great source for this animal but it came from other regions as well in the form of a gift or tribute. The kings of Gujarat on the west coast of India like Baroda, Bhavnagar, Kutch, Jamnagar and others had royal menageries overflowing with cheetahs. Bhavnagar boasted 32 African cheetahs early in the twentieth century. Some of these kingdoms also had African or foreign keepers and trainers to care for their cheetahs.
9. DNA tests do not distinguish cheetahs in India from those found elsewhere. Even if some consider the Iranian cheetah as a subspecies, the Indian cheetah could never have been one; if conclusive genetic studies are done on the remains of Indian cheetahs, it is likely that their genes will turn out to be much the same as any of the other subspecies and be highly inbred.
10. The cheetah was not a part of the Indian landscape and their numbers never added up, rather like the lion. Hundreds of cheetahs were imported into India in the twentieth century

and there is no reason to doubt that the same thing happened in the earlier centuries.

As we come to the end of the book, I would like to leave the reader with a closing thought, one that has to do with the status of the lion and cheetah in the country today. Where the small pride of docile Junagadh lions is concerned, the rulers of present-day Gujarat are no different from the Nawab of Junagadh. The state government regards these animals as the property of the state, just as the nawab did, and has blocked all attempts by other states, like Madhya Pradesh, to propagate the lion. The lions of Junagadh still depend heavily on livestock and the farmers who lose their livestock are frequently compensated. They get medicare and if injured, are treated and then released back into the forest. In a way, time stands still for them as history is repeated endlessly.

In the case of the cheetah, conservationists are calling for the Namibian subspecies to be introduced into India, either in Madhya Pradesh or Rajasthan, hence setting the stage for history to repeat itself in the case of the cheetah as well, with the animal being imported from Africa, just as it was for centuries. If imports had continued, I am sure the wealthy of today would have continued to have them as pets. Speaking for myself, I love both these animals and would rather watch them in their natural habitat—the wild open grasslands and bush country of Africa. They were never part of the Indian wild and, to my mind, will never be.

I believe that lions were imported into India a few thousand years ago as captive animals and bred in menageries and then propagated in private hunting parks. The Harappans seem ignorant of this animal and their sites reveal no clear representations of it. It is indicative of our cultures adopting the lion symbolism much later than many others.

ACKNOWLEDGEMENTS

The research for a book like this is highly complex, especially when you need to find visual support. I am indebted to the amazing museums of the world like The Victoria and Albert Museum, the British Museum, the British Library, the Metropolitan Museum of Art and others that enabled me to use what I needed for my research. This is an academic effort and the pictures used are to support my thesis. There are many friends who sent me pictures of lions and cheetahs from different sites and museums of the world that I would like to thank as well. I am especially grateful to Amanda Bright and Rohit Kaicker for their special inputs. I regret any errors that may have crept into these acknowledgements.

The Trustees of the British Museum:
Pictures on pages 30, 31, 77,114, 115 and the picture at the bottom on pages 116 and 139, taken by the author.
Pictures on pages 34, 50 and 67.
'Irresistible Imagery and Power: The Lion King':
Pictures 5 and 12.

The British Library Board:
Picture on page 21 taken by the author.
Pictures on pages 149 and 234.

John and Frank Craighead: The Archives of American Falconry publication, *Life with an Indian Prince*:
Pictures on pages 159, 175, 223, 224 and 225.

Private:
Pictures on pages 1, 22, 23, 25, 34, 40, 51, 57, 63, 64, 69, 83, 105, 106, 112, 125, 128, 134, 136, 137, 143, 145, 155, 160, 162, 165, 166, 167, 171, 178, 183, 185, 201, 203, 204, 207, 213, 217 and 293.
'Irresistible Imagery and Power: The Lion King':
Picture 1 (from the National Archaeological Museum of Athens), 2, 3 (from the Louvre Museum), 4, 6, 7, 8, 9, 11 (from the Egyptian museum at Turin), 15, 16, 17, 18, 19, 20, 21, 22, 23, 24, 25, 26, 27, 28, 29, 30 and 31.

Valmik Thapar:
Pictures on pages 21, 109 (with K. Sharma), 181, 185, 231 (left) and 293.
The picture on top on page 116.

The Vorderasiatisches Museum, Berlin:
Picture on top on page 36.

National Museum, New Delhi:
Picture on page 193.
Picture at the bottom on page 36.

Shaikh Usman Tirandaz (Paintings and Sketches Based on Mughal Art between the Sixteenth and Nineteenth Century CE):
Pictures on pages 9, 83, 96, 113, 129, 133, 157, 237.

Picture on top on page 110.

Karen Knorr, courtesy Tasveer©:
Pictures on page 5.

The Collection of Prince Sadruddin Aga Khan, Aga Khan Museum:
Picture on page 191 (from *In Danger*, published by the Ranthambhore Foundation).
Picture on top on page 194.

The Metropolitan Museum of Art, New York:
Pictures on pages 52, 53 and 89.

Victoria and Albert Museum, London:
Picture on the back cover.
Pictures on pages 6, 10, 66, 71, 72, 75, 90, 99, 101, 121, 187, 188, 197, 198 and 227.
Pictures at the bottom on pages 110 and 194.
'Irresistible Imagery and Power: The Lion King':
Pictures 13 and 14.

Bhushan Pandya\Sanctuary Photo Library:
Picture on the right on page 231.

Khuda Bakhsh Library, Patna:
Picture on page 97.

Supplied by Royal Collection Trust/© Her Majesty Queen Elizabeth II 2013:
Picture on page 122.

Istanbul Archaeological Museum (Author's Pictures):
Pictures on pages 2, 61 and 77.
'Irresistible Imagery and Power: The Lion King':
Picture 10.

The Trustees of the Chester Beatty Library, Dublin:
Picture on page 218.

NOTES

PROLOGUE

13 **He went on to state:** Captain Thomas Williamson, *Oriental Field Sports* (Naval & Military Press).

13 **'Those who deny':** Thomas Pennant, *The View of Hindoostan* 2 (London: Printed by Henry Hughs), p. 185.

13 **'The Hunting Leopard (Cynalurus jubatus) has now':** A. A. Dunbar Brander, *Wild Animals in Central India* (London: Edward Arnold & Co.), p. 273.

14 **As I researched into the:** Erwin Neumayer, *Lines on Stone: The Prehistoric Rock Art of India* (New Delhi: Manohar).

17 **'[The] physical traits':** Stephen J. O'Brien, *Tears of the Cheetah: And Other Tales from the Genetic Frontier* (New York: Thomas Dunne Books), p. 44.

17 **'virtually zero genetic diversity':** Ibid., p. 46.

17 **'The Gir lion males':** Ibid., p. 47.

18 **'When we considered':** Ibid., p. 51.

18 **'Even if this was':** Personal communication with historian Romila Thapar.

19 **'Over its natural range':** Ross Barnet, Nobuyuki Yamaguchi, Ian Barnes and Alan Cooper, 'The origin, current diversity and future conservation of the modern lion (*Panthera leo*)' (Proceedings of the Royal Society: Biological Sciences, 273, no. 1598, 7 September 2006).

19 **'Modern lions are':** L. Bartosiewicz, 'A Lion's Share of Attention: Archaeozoology and the Historical Record', *Acta Archaeologica* 60, no.1 (Budapest, Hungary: Institute of Archaeological Sciences, Eötvös Loránd University Múzeum, June 2009).

20 **'population bottleneck':** Stephen J. O'Brien, *Tears of the Cheetah: And Other Tales from the Genetic Frontier* (New York: Thomas Dunne Books), p. 28.

THE LION: FROM PRIDE TO METAPHOR

27 **The Chauvet-Pont-d'Arc Cave:** Jean Clottes, 'Chauvet Cave (ca. 30,000 BC)', *Heilbrunn Timeline of Art History* (New York: The Metropolitan Museum of Art, October 2002).

27 **As to why they were:** Sir James Frazer, *The Golden Bough: A Study in Magic and Religion* 1 (New York: Macmillan and Co. Ltd.); Joseph Campbell, *The Hero with a Thousand Faces* (New York: Pantheon Books).

28 **Animal herders were:** I. Schapera, 'A Native Lion Hunt in the Kalahari Desert', *Man* 32 (Royal Anthropological Institute of Great Britain and Ireland, December 1932), pp. 278-279.

28 **What is striking is:** W. M. Flinders Petrie, *The Making of Egypt* (London: Sheldon Press), chapter 7.

29 **In Egypt, the lion headed goddess:** Veronica Ions, *Egyptian Mythology* (London: The Hamlyn Publishing Group Ltd.), p. 86.

29 **That lion hunting was required:** I. E. S. Edwards, C. J. Gadd and N. G. Hammond, *Cambridge Ancient History* 1 (Cambridge University Press), p. 7; I. E. S. Edwards, C. J. Gadd, N. G. Hammond and E. Solleberger, *Cambridge Ancient History* 2, no. 1 (Cambridge University Press), pp. 219, 333.

29 **The hunters, mainly kings:** J. H. Breasted, *A History of Egypt* (London: Hodder & Stoughton Limited), pp. 350-351.

30 **In Mesopotamia:** Lions were spotted in Mesopotamia in the nineteenth century living in the shrubs along the banks of rivers; similarly, tigers were seen in the galleries of the Indus River as late as the nineteenth century.

30 **Once there, the animals were protected:** I. E. S. Edwards, C. J. Gadd, N. G. Hammond and E. Solleberger, *Cambridge Ancient History* 2, no. 1 (Cambridge University Press), p. 219.

30 **Tiglath-Pileser I claims:** I. E. S. Edwards, C. J. Gadd, N. G. Hammond and E. Solleberger, *Cambridge Ancient History* 2, no. 2 (Cambridge University Press), p. 463; W. M. Flinders Petrie, *The Making of Egypt* (New York: Sheldon Press), pp. 65-66.

30 **The different stages:** Julian Reade, *Assyrian Sculpture* (London: British Museum Press), p. 72.

31 **The procedures in the hunting parks:** A. L. Oppenheim, *Ancient Mesopotamia: Portrait of a Dead Civilization* (Chicago: University of Chicago Press), p. 46.

31 **In Greece, the famous gate:** I. E. S. Edwards, C. J. Gadd, N. G. Hammond and E. Solleberger, *Cambridge Ancient History* 2, no. 2 (Cambridge University Press), pp. 864, 874.

32 **The family of Alexander:** As did the ancient Sinhala royal family of Sri Lanka, even though they had probably never seen a lion.

32 **Further west, Etruscan art in Italy:** W. Llewellyn Brown, *The Etruscan Lion* (Oxford: Clarendon Press).

32 **In the neighbourhood of the eastern:** Deirdre Jackson, *Lion* (London: Reaktion Books), p. 115.

32 **Ptolemy II Philadelphus of Egypt:** Romila Thapar, *Asoka and the Decline of the Mauryas,* Third ed. (New Delhi: Oxford University Press, 2012), pp. 167-168.

33 **However, no mention:** One view holds that the idea of the Sarnath pillar of addorsed lions was borrowed from Egypt, but both the form and the concept are so dissimilar that such a borrowing is untenable.

33 **They were more likely a gift:** Andrew Nichols, *Ctesias: On India* (London: Bristol Classical Press, 2011), p. 100; Edward H. Schafer, *The Golden Peaches of Samarkand: A Study of T'ang Exotics* (Berkeley: University of California Press), p. 97.

33 **Maritime contacts:** F. De Romanis and A. Tchernia (eds.), *Crossings: Early*

Mediterranean Contacts with India (New Delhi: Manohar); J. K. Anderson, *Hunting in the Ancient World* (Berkeley: University of California Press), pp. 80-102.

33 **The Deccan:** E. H. Warmington, *The Commerce between the Roman Empire and India* (London: Curzon Press; New York: Octagon Books).

34 **Eastwards in Iran:** Ilya Gershevitch, *Cambridge History of Iran* 2 (Cambridge University Press), pp. 417-418.

35 **But the prize object:** John Curtis, *The Oxus Treasure (Objects in Focus)* (London: British Museum Press).

35 **'I am Darius the king...':** Ilya Gershevitch, *Cambridge History of Iran* 2 (Cambridge University Press), p. 725.

36 **The lion had a much stronger presence:** The evidence for Hellenistic forms is limited during this period in West Asia. V. Karageorghos (ed.), *The Greeks Beyond the Aegean: From Marseilles to Bactria* (New York: Alexander S. Onassis Foundation).

37 **It refers to the thunderous roar:** *Rig Veda*, 1.64.8, 1.154.2, 10.160.2, 3.26.5.

37 **A possible explanation:** This area in current archaeology is sometimes referred to as the Bactria-Margiana Archaeological Complex (BMAC), and is associated by some scholars with the earliest Indo-Aryan and Old Iranian languages (although some would prefer to place the original home in Anatolia).

38 **Unlike tigers:** Personal communication with Valmik Thapar.

38 **The tiger is mentioned:** *Atharva Veda*, 4.36.6, 5.20, 5.21, 8.7.15.

38 **Yet, when the forest:** *Mahabharata*, Adi Parva.

39 **The text—called Indika:** Andrew Nichols, *Ctesias: On India* (London: Bristol Classical Press), pp. 48-49, 60-63.

39 **'natives kill[ed] them':** Ibid.

40 **the 'Cakkavatti Sutta':** *Digha Nikaya*, 'Cakkavatti Sutta'; T. W. and trans. C. A. F. Rhys Davids, *Dialogues of the Buddha* 3 (London: The Pali Text Society).

42 **There are, however:** A. K. Coomaraswamy, *History of Indonesian and Indian Art*, (New York), pp. 11-18.

43 **I reckon Indian dogs:** J. W. McCrindle, *The Invasion of India by Alexander the Great* (London: Kessinger Publishing), pp. 363-364; Strabo, *Geography*, 15.1.31.

44 **And in the processions:** Strabo, *Geography*, 15.1.69.

44 **But there is no reference to a lion hunt:** Arrian, *Indika*, pp. 13-15, 18. J. W. McCrindle, *The Invasion of India by Alexander the Great* (London), pp. 261-63, 341.

44 **infested with wild tigers:** *Natural History* 6, p.73.

44 **Pliny also refers to:** E. H. Warmington, *The Commerce between the Roman Empire and India* (Delhi: Vikas Publishing House), pp. 147-148.

44 **Aelian, writing in:** Aelian, III. C. xxvi in J. W. McCrindle, *Ancient India as Described in Classical Literature* (Amsterdam), p. 149.

45 **As compared to the graphic evidence:** J. F. Nott, *Wild Animals: Photographed and Described* (London: Sampson Low, Marston, Searle & Rivington); F. E. Zeuner, *A History of Domesticated Animals* (London: Harper & Row); quoted in Divyabhanusinh, *The End of a Trail: The Cheetah in India* (New Delhi: Oxford

University Press), p. 15.

45 **The *Manasollasa*:** Divyabhanusinh, *The End of a Trail: The Cheetah in India* (New Delhi: Oxford University Press), p. 27.

47 **The emperor Ashoka writes:** Major Rock Edict VIII. J. Bloch, *Les Inscriptions d'Asoka* (Paris), p. 111.

47 **Buddhists believed:** J. S. Strong, *The Legend of King Asoka* (New Jersey: Princeton University Press), pp. 233-234, 285-286.

47 **The animals mentioned:** *Arthashastra*, 2.17.13; R. P. Kangle, *The Kautilya Arthasastra* (Mumbai: University of Bombay).

47 **...An animal park:** Ibid., 2.2.3.

48 **This again would not qualify:** Ibid.

48 **Another demarcated area:** Ibid., 2.26.1.

49 **In the second century CE:** T. T. Allsen, *The Royal Hunt in Eurasian History* (Philadelphia: University of Pennsylvania Press).

49 **The concept of the forest:** Romila Thapar, 'Perceiving the Forest: Early India', *Studies in History* 17, no. 1 (Sage), pp. 1-16.

51 **Kalidasa, in one of the opening verses:** *Kumarasambhava*, 1.6.

51 **The lion capitals:** J. Williams, 'A recent Asokan Capital and the Gupta attitude towards the past', *Artibus Asiae* XXXV, pp. 225-40; J. Williams, *The Art of Gupta India* (Princeton).

52 **However, Narasimha, not widely worshipped:** R. C. Hazra, *Studies in the Upapuranas* 2 (Calcutta), pp. 90, 219-266.

52 **The Narasimha image:** Deborah A. Soifer, *The Myth of Narasimha and Vamana: Two Avatars in a Cosmological Perspective* (Albany: SUNY Press), p. 73.

52 **Another myth mentions:** J. N. Banerjee, *The Development of Hindu Iconography* (Calcutta: University of Calcutta), pp. 231, 488.

52 **The man-lion also takes shape:** R. Deekshitar, 'Discovering the Anthropomorphic Lion in Indian Art', *Marg* 55, no. 4 (June 2004).

54 **The lion is either found:** C. Shivaramamurti, *The Art of India* (New York: Abrams), pp. 440, 516, 690.

54 **The ultimate fantasy:** P. Mitter, *Indian Art* (Oxford: Oxford University Press), pp. 36, 56, 64.

58 **Chinese Buddhist pilgrims:** S. Beal, *Travels of Fah-Hian and Sung-Yun* (London: Trübner & Co.), p. 199.

58 **Bringing horses from Central Asia:** The Pehowa inscription from the temple of Garibnath, *Epigraphia Indica* 1 (Archaeological Survey of India), pp. 184-190.

59 **These animals find their way:** E. H. Warmington, *The Commerce between the Roman Empire and India* (Delhi: Vikas Publishing House), pp. 146-148.

59 **That some were hunted and trapped:** I. E. S. Edwards, C. J. Gadd and N. G. Hammond, *Cambridge Ancient History* 1 (Cambridge University Press), pp. 207-208.

60 **Augustus received:** E. H. Warmington, *The Commerce between the Roman Empire and India* (Delhi: Vikas Publishing House), pp. 35-37.

60 **Peacocks and parrots:** Cassius Dio, *Roman History*, 68.15; Eusebius of Caesarea, *Life of Constantine* IV, p. 50.

60 **Tigers by contrast:** This is a point that Valmik Thapar has been emphasizing for some time.

ANIMALS OF THE CHASE

64 **The Austrian painter:** Wenzel Peter (1745-1829). This painting currently hangs at the Pinacoteca, the art gallery of the Vatican.

64 ***Book of the Dead* of Ani:** From the British Museum, London. See, Divyabhanusinh, *The Story of Asia's Lions* (Bombay: Marg Publications).

67 **The lion had a long history:** Yaghūth watches over the tribe, protecting them from evil and from their enemies. He represents brute strength and power, and is shown as a lion to demonstrate his violent power and aggression. As a war god, he is the patron of warriors, marching with them into battle and granting them victory.

67 **...Then they were taken down:** Hugh Kennedy, *When Baghdad Ruled the Muslim World* (Cambridge: Da Capo Press).

68 **...hunting with cheetahs:** A. L. Basham, *The Wonder that was India: A Survey of History and Culture of the Indian Sub-Continent Before the Coming of the Muslims* (London: Sidgwick & Jackson).

68 **They ask you:** Surat Al-M'idah.

69 **'two to three thousand':** Khaliq Ahmad Nizami, *Royalty in Medieval India* (Delhi: Munshiram Manoharlal Publishers Pvt. Ltd.), p. 97. Also see, Divyabhanusinh, *The Story of Asia's Lions* (Bombay: Marg Publications), p. 82.

69 **He took great pleasure:** Ziauddin Barni, *Tarikh-i-Firoz Shahi.*

69 **twenty kos:** Twenty kos in the Mughal measure of the kos would be approximately 82 kilometres. This would certainly encompass the hunting grounds of both Tughlaq and Mughal monarchs in proceeding centuries. What we see here is the emergence of present-day Haryana around the environs of Delhi as a favoured hunting area for successive monarchs.

70 **'It may well be':** Divyabhanusinh, *The End of a Trail: The Cheetah in India* (Banyan Books).

70 **The chase of the deer:** Ziauddin Barni, *Tarikh-i-Firoz Shahi*. Shams Siraj 'Afif. *Tenth Mukaddamah: Hunting Excursions.*

73 **Clearly, not all of these:** But compare the following description with one of Ptolemic Egypt, which describes the zoo built in the Museum of Alexandria by Ptolemy I (367-283 BCE) and Ptolemy II (309-246 BCE) that exhibited 'a variety of exotic species, such as elephants, antelope, camels, parrots, leopards, cheetahs, a chimpanzee, twenty-four "great" lions, lynxes, Indian and African buffaloes, a rhinoceros, a polar bear and a 45-foot long python.' Linda Kalof, *Looking at Animals in Human History* (London: Reaktion Books).

73 **Since the Monarch:** Shams Siraj 'Afif, *Tarikh-i-Firoz Shahi.*

73 **Of the single class:** Ibid.

73 **'Indeed, during his reign':** Ibid.

74 **The Sultán...:** Shams Siraj 'Afif, *Tarikh-i-Firoz Shahi, Eleventh Mukaddamah: Arrival of Sultán Fíroz at Hánsí.*

74 **In fact, so engaged:** 'Emperor Feroz Shah Tughlaq constructed the Western Yamuna Canal (WYC) to divert water to the hunting grounds in Hansi-Safidon area in Haryana in 1355 AD.' See, Prof. S. K. Jain, P. K. Agarwal and V. P. Singh, *Hydrology and Water Resources in India* (Netherlands: Springer.

74 **'In old times lions':** *Karnal District Gazetteer.*

THE SHAHENSHAH'S SHIKAR

77 **The special powers**: T. T. Allsen, *The Royal Hunt in Eurasian History* (Philadelphia: University of Pennsylvania Press), p. 10.

78 **During the march:** Timur, *Mulfazat Timury*, Trans. Charles Stewart.

78 **...the pleasantly-situated:** E. G. Browne, *A Year Amongst the Persians: Impressions as to the Life, Character, & Thought of the People of Persia,* (Cambridge: Cambridge University Press).

79 **the Persian lion (shir):** 'I mention this chiefly because this word, mispronounced *sher* (like English "share"), is applied in India to the tiger, which animal is properly termed babr in Persian, as stated in the text.' The note is the author's, and highlights once more the generic nature of the term *sher* which in India has been applied interchangeably to both the lion and the tiger.

79 **We were struck with:** S. A. Reis, *Mira'at ul Memalik (The Mirror of Countries).*

79 **We were obliged to fight:** Ibid.

80 **'...was a thriving town':** J. Marozzi, *Tamerlane: Sword of Islam, Conqueror of the World* (London: HarperCollins).

80 **Given its location, Turkestan:** The Muscovite Tsar, Boris Godunov, whose reign overlapped the last seven years of Akbar's reign, kept a lion in his menagerie. Early in the eighth century, when the Arab military was cracking down on the non-Muslim rulers of Turkestan, they turned to Tang China for support. According to Professor Allsen, 'This took the form of a series of diplomatic/tribute missions that presented, among other items, hunting cats to the Tang court on the assumption that the court would be delighted to receive them and hereby look more favourably on requests for military aid.' Not only did they exchange cheetahs, but also their keepers 'and it is this combination that facilitates and encourages uniformity in material culture.' T. T. Allsen, *The Royal Hunt in Eurasian History* (Philadelphia: University of Pennsylvania Press), p. 272.

80 **'He brought with him':** J. Marozzi, *Tamerlane: Sword of Islam, Conqueror of the World* (London: HarperCollins).

81 **This was the first time:** Abu'l-Fazl, *The Akbarnama,* trans. H. Beveridge, 3.

82 **The first hypothesis which:** The *Iqbalnama* of Jahangir's reign puts the figure of captive cheetahs at Akbar's court at 9,000, which is either a mistake or a classic case of Mughal hyperbole. Both Abu'l-Fazl and Jahangir state that Akbar collected 1000 cheetahs in his lifetime. This figure indicates the total number of cheetahs in the royal menageries, in the royal game reserves and on the

emperor's person when he travelled. Even 1,000 cheetahs is a fantastic figure; it suggests the emperor or his scouts were acquiring 15 cheetahs a year, every year for the duration of Akbar's life.

83 **'during Akbar's time':** Divyabhanusinh, *The End of a Trail: The Cheetah in India* (New Delhi: Banyan Books), p. 48.

83 **'in His Majesty's park':** Abu'l-Fazl, *Ain-i-Akbari,* trans. H. Jarrett. The *Ain-i-Akbari* further divides the cheetah into three categories, which were a) Horsed cheetahs, requiring three keepers for each and three horses for every pair; b) Carted cheetahs, requiring two keepers for each; c) Wild/In training cheetahs. The *Ain-i-Akbari* clearly states that the latter were not trained and were likely to escape.

83 **Once in the field:** Ibid.

84 **[Basing my calculations] on:** From Raghu. S. Chundawat's correspondence with the author.

84 **Therefore, to sustain themselves:** Compare this to the cheetah numbers in the collection of Qutb-ud-din Mubarak (1316-1320), mentioned in the previous chapter. His 2,000-3,000 'hunting-panthers' would automatically require a minimum black buck population of 8.46 million in the wild.

85 **On the first day:** Jahangir, *Tuzuk-i-Jahangiri* or *The Memoirs of Jahangir,* Vol. II, ed. H. Beveridge and trans. A. Rogers (Ghazipur), p. 70.

85 **When news came:** Ibid., p. 83. Compare this with a similar translocation in Vol I, p. 204.

86 **Barely three years after:** Ibid., p. 109.

87 **...no less than seven**: Abu'l-Fazl, *The Akbarnama* 3, trans. H. Beveridge.

88 **...his grandfather:** Shah Nawaz Khan, *Ma'athir-ul-Umara* 1, trans. H. Beveridge (Calcutta: Asiatic Society of Bengal), p. 261.

88 **For instance, two of:** Abu'l-Fazl, *The Akbarnama* 3, trans. H. Beveridge, p. 132.

88 **Annemarie Schimmel, in her book:** A. Schimmel, *The Empire of the Great Mughals: History, Art and Culture* (London: Reaktion Books).

91 **legendary Citr Najan at Akbar's:** Possibly Cītrānga, as Divyabhanusinh points out, or even Citranjan.

91 **'He also ordered that':** Abu'l-Fazl, *The Akbarnama,* trans. H. Beveridge.

91 **'His servants, fully equipped':** Ibid.

91 **Foreign dignitaries, his own nobles:** M. A. Alvi and A. Rahman, *Jahangir the Naturalist* (New Delhi: The National Institute of Sciences of India).

92 **'...to go to the port':** Jahangir, *Tuzuk-i-Jahangiri or The Memoirs of Jahangir,* ed. H. Beveridge and trans. A. Rogers (Ghazipur).

92 **Yet it was important enough:** The Mughals also imported elephants from Aceh, in Sumatra and received elephants as tribute from the Sultans of Golconda who, in their turn, had imported the beasts from Pegu, Ceylon and Siam. See Allsen, *The Royal Hunt in Eurasian History,* p. 252. The *Tuzuk-i-Jahangiri* also revels in the arrival of an exotic bird from Sumatra. Jahangir writes, 'In these days, they brought a bird from the country of Zirbād which was coloured like a parrot, but had a smaller body. One of its peculiarities is that it lays hold with its

feet of the branch or perch on which they may have placed it and then makes a somersault, and remains in this position all night and whispers to itself.'

92 **Muqarrab Khan presented:** Jahangir, *Tuzuk-i-Jahangiri* or *The Memoirs of Jahangir*, ed. H. Beveridge and trans. A. Rogers (Ghazipur, 1863). The editor refers to the *Akbarnama* (Vol. II, p. 315) and notes that 'it, too, was an African elephant'.

92 **'On the 12th the offering':** Ibid. Here the editor indicates the Arabian dogs to be greyhounds but is unsure of the identity of the 'hunting animals'.

92 **'...the offering of Lachmi Chand':** Jahangir, *Tuzuk-i-Jahangiri* or *The Memoirs of Jahangir*, ed. H. Beveridge and trans. A. Rogers 2 (Ghazipur), p. 191.

92 **Raja Bir Singh Deo brought:** Ibid., Vol. I.

93 **Given that the sources understandably:** Anthony Cutler, 'Gifts and Gift Exchange as Aspects of the Byzantine, Arab, and Related Economies', *Dumbarton Oaks Papers* 55 (2001), pp. 247-278.

93 **On one occasion, he records:** Jahangir, *Tuzuk-i-Jahangiri* or *The Memoirs of Jahangir*, Ed. H. Beveridge and trans. A. Rogers, 1.

94 **...ten huge elephants**: Munshi, '*Shah Abbas*', 2.

94 **'I went out to hunt tigers':** Jahangir, *Tuzuk-i-Jahangiri* or *The Memoirs of Jahangir* 1, ed. H. Beveridge and trans. A. Rogers. The editor gives the translation for 'black ear' or *siyaghosh*, as lynx, though more properly this is the caracal.

94 **On the top of this doorway**: Gonzalez De Clavijo. See, J. Marozzi, *Tamerlane: Sword of Islam, Conqueror of the World* (London: HarperCollins).

95 **The Persian lion is now extinct:** Jaffa, N. A. 'The Asiatic or Persian lion (*Panthera leo persica*, Meyer 1826) in Palestine and the Arabian and Islamic Region', *Gazelle: The Palestinian Biological Bulletin*, no. 58 (October, 2006), pp. 1-13.

95 **The Tutinama, commissioned:** S. P. Verma, 'Painting under Akbar as Narrative Art', *Akbar and His India*, ed. I. Habib (New Delhi: Oxford University Press India, 1997). 'The donkey wearing the skin of the tiger...' (MS folio 207r). 'The illustrative miniature...shows a donkey braying clad in tiger's skin [lion's skin in the text]...'

95 **During the Mughal reign:** Ibid.

96 **'At the beginning of this year':** N. A. Bakhshi, 'Tabaqat-i-Akbari', *The History of India as Told by its Own Historians*, ed. H. M. Elliot and J. Dowson (Low Price Publications), p. 329.

98 **'It was in all probability':** Ibid.

98 **At this time I saw:** Jahangir, *Tuzuk-i-Jahangiri* or *The Memoirs of Jahangir*, ed. H. Beveridge and trans. A. Rogers, Ghazipur. 2, p. 201.

98 **'It was like a tiger':** Ibid.

99 **You must know:** H. a.-D. Mirza, *Baznama-i-Nasiri: A Persian Treatise on Falconry*, trans. L. C. Phillott (London: Bernard Quaritch).

100 **The memoirs of Jahangir refer:** The citations for this are numerous. For instance, see Jahangir, *Tuzuk-i-Jahangiri* or *The Memoirs of Jahangir*, ed. H. Beveridge and trans. A. Rogers, Ghazipur, Vol. II, pp. 12, 27, 53-54, 58-60, 115, 246.

101 **Between 1690 and 1830:** R. H. Grove, *Green Imperialism: Colonial Expansion, Tropical Island Edens and the Origins of Environmentalism 1600-1860* (Cambridge: Cambridge University Press), p. 387.

102 **'as proof of his justness':** Niccolao Manucci, *Storia do Mogor* 2.

102 **'...Shah Abbas repaid':** Ibid.

IN THE COURTS OF EMPERORS

106 **The tameness of these animals:** F. Bernier, *Travels in the Mogul Empire A.D. 1656-1668,* ed. V. A. Smith and trans. A. Constable (Asian Educational Services reprint), p. 375.

107 **Moreover, experts have testified:** C. Guggisberg, *Simba, the Life of the Lion* (Cape Town: Howard Timmins), p. 42.

108 **Ali Akbar Khitai:** T. T. Allsen, *The Royal Hunt in Eurasian History* (Philadelphia: University of Pennsylvania Press), p. 236.

111 **It was because men:** John Hampden Porter, *Wild Beasts* (Charles Scribner's Sons).

111 **'In ancient Greece the priests':** C. Guggisberg, *Simba, the Life of the Lion* (Cape Town: Howard Timmins), p. 241.

111 **Tame lions were kept:** Ibid.

111 **Because of their long association:** Daniel Hahn, *The Tower Menagerie* (Pocket Books), p. 93.

112 **This was in 917 BCE:** Deirdre Jackson, *Lion* (London: Reaktion Books), p. 74.

112 **'Persian influence in architecture':** M. A. Rashid and Reuben David, *The Asiatic Lion* (Department of Environment, Government of India), p. 20.

112 **'that Egypt inspired':** Saryu Doshi, *India and Egypt* (New Delhi: Marg Publications). Remember that tame lions were kept at Leontopolis, the lion city in the Nile delta and even near modern-day Cairo to propitiate them.

113 **In 1592, a lion and lioness:** Daniel Hahn, *The Tower Menagerie* (Pocket Books), p. 110.

113 **Hunting parks blossomed across:** T. T. Allsen, *The Royal Hunt in Eurasian History* (Philadelphia: University of Pennsylvania Press), pp. 40, 41. In a poem by Einhard that talks about Charlemagne having a pleasure park that was 'enriched by many walls' near Aachen, it is told that the monarch would often go hunting there armed with dogs and arrows, 'laying low multitude of antlered stags beneath the black trees'. This description calls to mind the typical depictions of Iranian paradise.

114 **Paralysed by arrows:** Deirdre Jackson, *Lion* (London: Reaktion Books), pp. 159-160.

114 **The Moghul Emperors of:** C. Guggisberg, *Simba, the Life of the Lion* (Cape Town: Howard Timmins), p. 170.

114 **'fantastic massacres':** Ibid.

114 **'Indian princes may occasionally':** Ibid.

115 **'A small state in Turkestan':** T. T. Allsen, *The Royal Hunt in Eurasian History* (Philadelphia: University of Pennsylvania), p. 235.

115 **This is where Alexander:** Aldo Ricci, *The Travels of Marco Polo* (New Delhi: Rupa Publications India Pvt. Ltd.).

115 **Ibn Batuta, who travelled through:** H. Gibb, *The Travels of Ibn Batuta AD 1325-1354*, 3 (Cambridge: Cambridge University Press), pp. 597, 727.

115 **Within the lande there are:** Jan Huyghen van Linschoten, Arthur Coke Burnell and Pieter Anton Tiele, *The Voyage of John Huyghen van Linschoten to the East Indies: from the old English translation of 1598: the first book, containing his description of the East* (London: Printed for the Hakluyt Society), p. 305.

116 **In the yeare 1581:** Ibid., p. 10.

116 **Of tigers there are a vast number:** Albert Gray, *The Voyage of Francois Pyrard of Laval* 1 (New Delhi: Asian Educational Services), pp. 346-347. Laval, like van Linschoten, also provides information on the trade in exotic animals. He says this trade took place between ports the Portuguese controlled from Muzambique (modern-day Mozambique) to Mumbasa (Mombasa) and Ormuz in the Persian Gulf. Laval describes a boat that was pillaged and sunk: 'The pirates set fire to the ship, which contained a good number of Persian and Ormus horses.' Ormus is said to have supplied live lions, giraffes and leopards to Chinese traders like Zheng He in the fifteenth century and must have been strategic for animal imports into India.

116 **He mainly focuses on:** William Foster, *The Voyage of Thomas Best to the East Indies 1612-1614* (New Delhi: Asian Educational Services).

117 **'none but the king can hunt the lion':** J. C. Daniel, *A Century of Natural History* (Mumbai: Bombay Natural History Society).

117 **'Lions here are...feeble and cowardly':** Sir Thomas Roe and Dr John Fryer, *Travels in India in the seventeenth century: reprinted from the Calcutta Weekly Englishman* (London: Trübner & Co.), p. 238.

117 **At night the Kings custome:** Sir William Foster, *Early Travels in India* (Oxford: Oxford University Press), p. 111.

118 **Of Elephants the King keepeth:** Thomas Coryate, *Coryatt's Crudities* 8.

118 **There were also:** Sir William Foster, *Early Travels in India* (Oxford: Oxford University Press), p. 17.

118 **He was a great experimentalist:** Jahangir, *Tuzuk-i-Jahangiri* or *The Memoirs of Jahangir*, ed. H. Beveridge and trans. A. Rogers (Ghazipur).

119 **The rest of the daie:** William Foster, *The Journal of John Jourdain, 1608-1617, Describing his Experiences in Arabia, India, and the Malay Archipelago* (Cambridge: The Hakluyt Society), p. 159.

120 **For the lions to be safe:** Vincent A. Smith, *The Early History of India*; Jean-Baptiste Tavernier, *Travels in India* 1 (Macmillan & Co. Ltd.), p. 97.

123 **The elephants were treated well:** Valérie Berinstain, *Mughal India: Splendours of the Peacock Throne*, trans. Paul G. Bahn (London: Thames and Hudson).

123 **On the 4th the huntsmen:** Jahangir, *Tuzuk-i-Jahangiri or The Memoirs of Jahangir*, ed. H. Beveridge and trans. A. Rogers, Ghazipur.

123 **Such scenes bring to mind:** Stuart Cary Welch, *Gods, Kings and Tigers: The Art of Kotah* (Prestel).

124 **The paintings are full of clues**: Zaheda Khanam, *Birds and Animals in Mughal Miniature Paintings* (New Delhi: DK Printworld).

125 **As a preliminary step:** F. Bernier, *Travels in the Mogul Empire AD 1656-1668,* ed.V. A. Smith and trans A. Constable (Oxford: Oxford University Press), p. 378.

126 **'Hunting lions was...a mechanism':** T. T. Allsen, *The Royal Hunt in Eurasian History* (Philadelphia: University of Pennsylvania Press), p. 135.

126 **'hunted the lion, the elephant':** Thomas Pennant, *The View of Hindoostan,* Vol. 2, (London: Henry Hughs).

126 **'In many parts of Northern':** Stephen P. Blake, *Shahjahanabad: The Sovereign City in Mughal India, 1639-1739* (Cambridge: Cambridge University Press), p. 145.

127 **'Beasts of prey were taken':** William Jardine on Baron Cuvier's work, *Lions, Tigers, & c., & c.* (London: H. G. Bohn), pp. 62, 66. He goes on to state: 'By many of the Indian sovereigns, beasts of prey were kept to be hunted or to be tamed, were placed near the throne upon occasions of pomp—they were also much more frequently employed as the executioners of criminals…'

127 **Except in Kathiawar, lions are:** F. Bernier, *Travels in the Mogul Empire AD 1656-1668,* ed. V. A. Smith and trans. A. Constable (Oxford: Oxford University Press), p. 378.

127 **'[the] enraged animal leapt over the net'**: Ibid.

127 **'[a lion] was at length found':** Ibid.

127 **portentous of infinite evil:** F. Bernier, *Travels in the Mogul Empire AD 1656-1668,* ed.V. A. Smith and trans. A. Constable (London: Oxford University Press). See also, Loius Rousselet, *India and Its Native Princes: Travels in Central India and in the Presidencies of Bombay and Bengal* (Asian Educational Services). Loius Rousselet writes of the astrological elements of the hunt: 'The Guicowar was exceedingly superstitious. For several days we postponed our hunting parties because the astrologers had not been able to fix on a suitable day to commence them…and announce to the king with a meloncholy [sic] air, that the omens were not favourable…'

128 **'to see a lion brought':** S. N. Sen, Jean de Thévenot, Giovanni Francesco Gemelli Careri, and David E. Pingree's collection, *Travels of Jean De Thevenot and Gemelli Careri* (New Delhi: National Archives of India).

130 **In 1490, when Firoz Shah Tughlaq:** Stan Goron, 'The Habshi Sultans of Bengal', *African Elites in India – Habshi Amarat,* ed. Kenneth Robbins and John McLeod (Ahmedabad: Mapin), p. 133.

130 **'unchallenged masters':** Edward A. Alpers, 'Africans in India and the wider context of the Indian Ocean', *Sidis and Scholars: Essays on African Indians,* ed. Amy Catlin-Jairazbhoy and Edward A. Alpers (Trenton, NJ: Rainbow Publishers), pp. 27-41.

130 **A Mughal miniature painting:** *African Elites in India: Habshi Amarat,* ed. Kenneth Robbins and John McLeod (Ahmedabad: Mapin), p. 163.

130 **They knew the territories:** Lieut. Col. L. L. Fenton, *The Rifle in India: Being the Sporting Experiences of an Indian Officer* (London: W. Thacker and Co.).

130 **There is a widespread stereotype**: In an interview given to *Frontline* 27, no. 18, 27 August to 9 September 2005.

131 **They kill Lions for sport:** S. N. Sen, Jean de Thévenot, Giovanni Francesco Gemelli Careri, and David E. Pingree Collection, *Travels of Jean De Thevenot and Gemelli Careri* (New Delhi: National Archives of India).

THE LAST LIONS

135 **'As to lions there are none':** Captain Thomas Williamson, *Oriental Field Sports* (New York: printed for Edward Orme), p. 130.

135 **In the neighbourhood**: Thomas Pennant, *The View of Hindoostan* 2 (London: Printed by Henry Hughs).

136 **Allsen writes about:** T. T. Allsen, *The Royal Hunt in Eurasian History* (Philadelphia: University of Pennsylvania Press), p. 18.

136 **Having watched for:** James Forbes, *Oriental memoirs: A Narrative of Seventeen Years Residence in India 2, revised by his daughter, Countess de Montalembert* (London: Richard Bentley), p. 178.

138 **On a signal given by the Rajah:** Captain Basil Hall, *Travels in India, Ceylon and Borneo* (London: Routledge). See also, Louis Rousselet, *India and Its Native Princes: Travels in Central India and in the Presidencies of Bombay and Bengal* (New Delhi: Asian Educational Services), p. 115, where he states that (lions) were to be found along with tigers, panthers and bears in menageries. 'These creatures are kept under sheds, and merely attached to posts by long chains. The visitor is obliged to walk circumspectly and although the chains are strong, it is not very comfortable to be in the midst of so ferocious a company. A beautiful black panther was chained at the door, so that, in order to enable you to go in or out, it was necessary for a keeper to hold her back...in another building were the cheetahs and lynxes employed in the chase...in a pavillion attached to the menagerie are the falcons, hawks and buzzards trained for the pursuit of birds...'

139 **What is fascinating about the lion-buffalo fights:** William Daniell and Hobart Caunter, *The Oriental Annual, or Scenes in India* (London: Bull and Churton), p. 156.

140 **Organizing fights:** William John Jeremy Chubb, *The Lucknow Menagerie: Natural History Drawings from the Collection of Claude Martin (1735-1800).*

140 **'indeed the existence':** *The Asiatic Journal and Monthly Register for British India and It's Dependencies* 3, no. XV (London: March 1817), pp. 228-229. 'Exploits of a Lion Shooting Party of English Gentlemen at Baroda', *The Annual Register: A View of the History, Politicks and Literature of the Year...* 59, ed. Edmund Burke (London: Printed for Baldwin, Cradock, and Joy, 1818).

141 **'The place in which they were':** Ibid, p. 565.

141 **'with the bayonet':** Ibid, p. 566.

141 **'on being brought':** Ibid, p. 567.

141 **M. A. Rashid, a senior officer:** M. A. Rashid & Reuben David, *The Asiatic Lion* (Department of Environment, Government of India), p. 37. The above figure does not include improbable claims such as the one made by Colonel

Smith who boasted that he had shot 50-300 lions near Delhi!

141 **'It is curious that':** R. G. Burton, *A Book of Man Eaters* (London: Hutchinson & Co. Ltd., 1817; reprint by Mittal Publishers), p. 75.

141 **1847 – one in Damoh:** Ibid.

142 **There were Lions:** R. G. Burton, *The Book of the Tiger* (London: Hutchinson & Co. Ltd., 1824; reprint by Mittal Publishers), p. 278.

142 **Three shikaris arrived:** R. G. Burton, *A Book of Man Eaters* (London: Hutchinson & Co. Ltd., 1817; reprint by Mittal Publishers), p. 271.

142 **My last letter:** Victor Jacquemont, *Letters from India: Describing a Journey in India, Tibet, Lahore and Cashmere, During the Years 1828, 1829, 1830, 1831* (Karachi, New York: Oxford University Press), p. 17.

144 **James Tod:** James Tod, *Annals and Antiquities of Rajasthan* 3 (reprint by Motilal Banarsidass), p. 1749. Tod witnessed many massacres of wildlife in stage-managed hunts carried out on a scale that was beyond belief: '…as the hunters approached a bevy of animals, amongst which some black snouted hyeanas [sic] were seen…a slaughter commenced…after I went to my tents I found six camel loads of deer, of various kinds deposited…still it was an exhilarating scene; the confusion of the animals, their wild dismay at their compulsory association; the yells, shouts and din from four battalions of regulars…formed a chain from the summit of the mountain across the valley to the opposite heights…hunting excursions cost the state two lakhs or 20,000 pounds annually. The regents [sic] regular hunting establishment consisted of 25 carpenters, two hundred aherias, or huntsmen, and 500 occasional rangers.'

144 **mentions lions as being a part of the prize:** Percival Spear, *The Nabobs: A Study of the Social Life of the English in Eighteenth Century India* (London: Oxford University Press), p. 89. Even then, the tradition of 'netting' animals in private hunting areas continued. In this book, the historian quotes Sir Martin Hunter: 'At the hunts of the Nawab of Arcot, who attended "with a wonderfully large retinue", a net about a mile long was stretched outside of a jungle…made of very strong cord as thick as my finger.' When the Nawab arrived on his elephant, the 'poligars' would make a din using 'matchlocks, sounding horns and beating drums', driving the frightened animals to the edge of the enclosure where they were mercilessly shot. 'I was much surprised,' added Sir Martin, 'that they did not shoot one another; a regiment could not have kept up a more constant fire for nearly an hour.'

144. **In spite of the fact:** Layard L. Fenton, *The Rifle in India: Being the Sporting Experiences of an Indian Officer* (London: W. Thacker and Co.), p. 1.

146 **37 lions were shot:** Paul Joslin, *The Asiatic Lion: A Study of Ecology and Behaviour* (University of Edinburgh).

146 **miniscule amount when compared to:** Personal communication with Valmik Thapar.

146 **Mahesh Rangarajan believes:** Mahesh Rangarajan, *India's Wildlife History: An Introduction to the History of India's Wildlife, Culminating in the Present Crisis* (Delhi: Permanent Black, in association with Ranthambhore Foundation). Mahesh

Rangarajan is a historian and specializes in modern history. Presently, he is the director of the Jawaharlal Nehru Library and Museum.

146 **The lion was once:** R. G. Burton, *The Book of the Tiger* (London: Hutchinson & Co. Ltd., 1824; reprint by Mittal Publishers), p. 273.

146 **It is curious:** Ibid., p. 274.

146 **'Lions (*Felis leo*) were':** Jan Huyghen van Linschoten, Arthur Coke Burnell and Pieter Anton Tiele, *The Voyage of John Huyghen van Linschoten to the East Indies: from the old English translation of 1598: the first book, containing his description of the East* (London: Printed for the Hakluyt Society), p. 305.

147 **'The shikaris of Kotah':** Divyabhanusinh, *The Story of Asia's Lions* (Bombay: Marg Publications), p. 191.

147 **'ten lions were shot':** Ibid., p. 122.

147 **'They are young':** W. Cornwallis Harris, *Portraits of the Game and Wild Animals of South Africa* (South Africa: Galago Publishing).

148 **They did not always:** R. G. Burton, *A Book of Man-Eaters* (London: Hutchinson & Co. Ltd., 1817; reprint by Mittal Publishers, 1984), p. 75. In this connection, one should also mention an article in *The Bengal Sporting Magazine* of 1833, which printed an account of a lion hunt in Haryana, where the lion was killed in a 'tame way'.

148 **'While on the subject':** Sir David Davidson, *Memories of a Long Life* (David Douglas), p. 200.

150 **The Sahibs from Shahjehanpoor:** Reginald Heber, *Narratives of a journey through the upper provinces of India: From Calcutta to Bombay, 1824-1825, (with Notes upon Ceylon,) an Account of a Journey to Madras and the Southern Provinces, 1826, and Letters Written in India* (London: John Murray).

150 **It is difficult:** R. G. Burton, *The Book of the Tiger* (London: Hutchinson & Co. Ltd., 1824; reprint by Mittal Publishers), p. 270.

150 **Edward Archer:** Edward Caulfield Archer was a British missionary who traversed the upper reaches of India and the Himalayas in the first part of the nineteenth century.

150 **The eastern or lion door:** Edward Caulfield Archer, *Tours in Upper India, and in Parts of the Himalaya Mountains; With Accounts of the Courts of Native Princes, & c.* (London: Richard Bentley), p. 144.

152 **'the lower-class Europeans'; '[just] as the British':** Christine Brandon-Jones, 'Edward Blyth, Charles Darwin, and the Animal Trade in Nineteenth-Century India and Britain', *Journal of the History of Biology* 30, no. 2 (Netherlands: Kluwer Academic Publishers), pp. 145-178.

152 **The beginning of the:** Nigel Rothfels, *Savages and Beasts: The Birth of the Modern Zoo* (Baltimore: The Johns Hopkins University Press), pp. 44-50.

152 **'Without exception':** Karl Hagenbeck. Ibid.

153 **Those who have been:** Frank Finn, *Sterndale's Mammalia of India* (Bombay: Thacker & Co.).

154 **'the Indian and African species':** Sir Montague Gilbert Gerard, *Leaves from the Diaries of a Soldier and Sportsman during Twenty Years' Service in India,*

Afghanistan, Egypt and Other Countries, 1865-1885 (E. P. Dutton).

154 **A family of lions:** Jules Gerard, *Lion Hunting: Adventures and Exploits of Famous Hunters and Travellers* (Ward, Lock and Co.).

155 **The lion is now confined to:** Sir William Wilson Hunter, *The Indian Empire, its People, History, and Products* (London: Trübner & Co.).

155 **In 1893, when the Durbar:** Richard Lydekker, *The Game Animals Of India, Burma, Malaya, And Tibet: Being A New And Revised Edition Of The Great And Small Game Of India, Burma, And Tibet.*

156 **'[they] proved very scarce'**: Philip W. Sergeant, *The Ruler Of Baroda* (London: John Murray), p. 87.

156 **'The Indian lion appears to differ':** Charles Edward Mackintosh Russell, *Bullet and Shot in Indian Forest, Plain and Hill—With Hints to Beginners in Indian Shooting* (London: W. Thacker and Co.), p. 196.

THE AGE OF SLAUGHTER

160 **On the way back to Rewah:** Rev. Edward St Clair Weeden, *A Year with the Gaekwar Of Baroda* (Boston: Dana Estes and Co.), p. 206.

161 **It is the custom in India:** Louis Rousselet, *India and its Native Princes* (Delhi: B. R. Publishing Corporation), p. 374.

162 **After failing to bag any game:** John Zubrzycki, *The Last Nizam,* (Australia: Pan Macmillan), p. 101.

162 **The dazzling roster of aristocratic:** Ed. by Clark Worswick, *Princely India* (London: Thames & Hudson), p. 12.

163 **Being a royal cage:** Christopher Armstead, *Princely Pageant* (London: Thomas Harmsworth Publishing), p. 72.

163 **'the rigours of camp life':** Ibid., p. 73.

163 **Within the vehicle:** Sir Kenneth Samuel Fitz, *Twilight of the Maharajas* (London: John Murray), p. 151.

163 **'regarded his tigers as':** H. M. Bull and K. N. Haksar, *Madhav Rao Scindia of Gwalior – 1876-1925* (Gwalior: Alijah Darbar Press), pp. 240-241.

163 **'how it annoyed Scindia!':** Ibid.

163 **'to ensure docility':** Charles Allen and Sharada Dwivedi, *Lives of the Indian Princes* (Century Publishing, in association with the Taj Hotel Group).

164 **'the nadir of shikar'**: Ibid., p. 117.

164 **'Tigers are trapped for many reasons':** Peter Ryhiner, as told to Daniel P. Mannix, *The Wildest Game* (London: Cassell), p. 169.

164 **'of darker hue which verges on the red':** Madhav Rao Scindia, *A Guide to Tiger Shooting.*

164 **The Maharaja of Bikaner, Ganga Singh:** Ian Copland, *The Princes of India in the Endgame of the Empire 1917-1947* (Cambridge: Cambridge University Press), p. 11.

165 **On the day before his shoot:** Major-General D. K. Palit, *Musings and Memories* (New Delhi: Palit and Palit, in association with Lancer Publishers).

166 **Frenzied with thirst:** Maud Diver, *Royal India* (New York: D. Appleton &

Company), p. 84.

166 **'...it was awful':** Ann Morrow, *The Maharajas of India* (New Delhi: Srishti Publishers and Distributors), p. 145.

167 **According to Sir Kenneth Fitz:** Sir Kenneth Samuel Fitz, *Twilight of the Maharajas* (London: John Murray), p. 141.

167 **'would have the first shot':** Charles Allen and Sharada Dwivedi, *Lives of the Indian Princes* (Century Publishing, in association with the Taj Hotel Group), p. 113.

167 **'he...completed the construction':** Sir Kenneth Samuel Fitz, *Twilight of the Maharajas* (London: John Murray), p. 140.

167 **There must have been:** Ibid., p. 116.

168 **'the finest partridge shoot he'd had anywhere':** Ibid.

168 **Unfortunately two of the cubs:** Sudipta Mitra, *Gir Forest and the Saga of the Asiatic Lion* (New Delhi: Indus Publishing Company).

168 **Finally they were caught:** Asheem and Suvira Srivastav, *Asiatic Lion on the Brink* (Dehradun: Bishen Singh Mahendra Pal Singh), p. 109.

168 **The Maharaja used to tell:** Marshall Lord Birdwood, *Khaki and Gown* (London: Ward, Lock & Co.), p. 54.

169 **the habitat of African lions:** Moreover, they were tiger-infested, which would prove detrimental to the welfare of the lions. I have no idea how many times the Maharaja tried to introduce lions into his forests, but the following excerpts from a book, *Madhav Rao Scindia of Gwalior: 1876-1925* (Gwalior: Alijah Darbar Press) by H. M. Bull and K. N. Haksar show that it was an enduring obsession: 'The gift of certain lion cubs from Lord Kitchener and the purchase of some more from Mombasa aroused in his Highness a wish to introduce lions once more into the state preserves.' I believe the regions between Agra and Gwalior are areas in which lions have been bred since the Mughal era.

170 **'Owing to breeding':** Colonel Kesri Singh, *One Man and a Thousand Tigers* (London: Robert Hale Limited), p. 162.

171 **lions and tigers could not coexist:** Ibid., p. 115. In Shivpuri, wild tigers were frequently seen near the enclosures lions were kept in before they were released. Kesri Singh recalls: 'It was also near Shivpuri that I encountered two tigers together in the road in broad daylight. This was during the rainy season when tigers are more inclined than usual to move about by day...'

171 **On the 20th (April 1912):** Sir Guy Douglas Arthur Fleetwood Wilson, *Letters to Nobody, 1908-1913* (New York: Dutton), p. 141.

172 **All the lions were his property:** Asheem and Suvira Srivastav, *Asiatic Lion on the Brink* (Dehradun: Bishen Singh Mahendra Pal Singh), p. 56.

173 **I was surprised to notice:** Colonel Kesri Singh, *One Man and a Thousand Tigers* (London: Robert Hale Limited), p. 162.

173 **'In a short span':** Asheem and Suvira Srivastav, *Asiatic Lion on the Brink* (Dehradun: Bishen Singh Mahendra Pal Singh), p. 56.

173 **He came back and bolted:** Ibid., p. 68.

173 **'the inhabitants...tried':** Ibid., p. 69.

173 **'They were audacious enough':** Ibid., p. 102.

174 **The Gir foresters':** C. A. Kincaid, *Forty-four Years a Public Servant* (Edinburgh: William Blackwood and Sons), p. 97.

174 **'It is customary for the head':** John J. Craighead and Frank C. Craighead Jr., *Life with an Indian Prince* (Idaho: Archives of American Falconry), p. 172.

175 **Some findings about the:** Stephen O'Brien, *Tears Of The Cheetah* (New York: Thomas Dunne Books).

177 **The two chowkidars**: Kenneth Anderson, *Jungles Long Ago* (New Delhi: Rupa Publications India Pvt. Ltd. by arrangement with George Allen & Unwin Ltd.), p. 140.

177 **'The relationship between':** Kailash Sankhala, *Return of the Tiger* (New Delhi: Lustre Press), p. 148.

178 **Evidence of lions:** M. A. Rashid and Reuben David, *The Asiatic Lion* (New Delhi: Department of Environment, Government of India), p. 34.

180 **'Kings set aside':** Laura Betzig, *Hunting Kings* (University of Michigan, Sage Publications).

THE STORY OF THE CHEETAH

183 **He behaved as though:** T. Murray Smith, *The Nature of the Beast* (London: Jarrolds Publishers Limited).

184 **'which greatly facilitates':** Odette de Puigaudeau, *Wild Animals Of Africa* (The Central Press), p. 71.

184 **'the cheetah, as the fastest animal':** Mel Sunquist and Fiona Sunquist, *Wild Cats of the World* (Chicago: University of Chicago Press), p. 20.

185 **His findings were disputed:** John Platt, 'Asian cheetahs racing towards extinction', *Scientific American* (24 January 2011).

185 **According to some scientists:** Stephen O'Brien, in his book, *Tears of the Cheetah* (New York: Thomas Dunne Books, 2003) stated: '...smack on the date of the great mammal extinction events at the end of the Pleistocene. So whatever had decimated the saber-tooths, mastodons, and giant sloths had nearly claimed the cheetahs as well. That narrow escape from extinction left its mark of genetic uniformity on the survivors...'

186 **Divyabhanusinh has stated:** Divyabhanusinh, *The Story of Asia's Lions* (Bombay: Marg Publications), p. 123.

186 **Although cheetahs are not:** Babur, *Baburnama: Memoirs of Babur*, trans. Annette Beveridge (New Delhi: Oriental Books Reprint Co.).

186 **'...panthers are drawn':** J. S. Hoyland and S. N. Banerjee, *The Commentary of Father Monserrate S. J. on his journey to the court of Akbar* (New Delhi: Asian Educational Services), pp. 77-78.

186 **The *Ain-i-Akbari:*** From the *Ain-i-Akbari* by Abu'l-Fazl:
Righrii—The leopard lies concealed, and is shown the deer from a distance. The collar is then taken off, when the leopard, with perfect skill, will dash off, jumping from ambush to ambush till he catches the deer.

Muhari—The leopard is put in an ambush, having the wind towards himself. The cart is then taken away to the opposite direction. This perplexes the deer, when the leopard will suddenly make his way near it and catch it.

189 **According to Mahesh Rangarajan:** Mahesh Rangarajan, *India's Wildlife History: An Introduction to the History of India's Wildlife, Culminating in the Present Crisis* (Delhi: Permanent Black and Ranthambhore Foundation).

189 **One of the joy-increasing:** Akbar, *Akbarnama,* trans. H. Beveridge 8.

190 **Seventeenth-century French traveller:** Jean de Thévenot, *Travels of Monsieur De Thevenot into the Levant* (London: Faithorne). '…there are a great many forests around Ahmedabad where they take panthers for hunting, and the governor of the town causes them to be taught, that he may send them to the king. The governor suffers none to buy them but himself, and they whose care it is to tame them in the meidan where from time to time they stroak and make much of them, that they may accustom them to the sight of men.'

191 **The cheetah or hunting leopard:** Sir William Hunter, *The Indian Empire: Its People, History, and Products* (London: Trübner & Co.), p. 653.

192 **Henry Bevan:** Henry Bevan was a soldier and sportsman who spent 30 years in India.

192 **Of late the cheetahs:** Henry Bevan, *Thirty Years in India: or, A soldier's reminiscences of native and European life in the presidencies, from 1808 to 1838* 2 (P. Richardson), p. 272.

192 **The hunting of the antelope:** Henry Bevan, *Thirty Years in India: or, A soldier's reminiscences of native and European life in the presidencies, from 1808 to 1838* 1 (P. Richardson), p. 220.

195 **[First] five Elephants:** Sir Thomas Roe and Dr John Fryer, *Travels in India in the seventeenth century: reprinted from Calcutta Weekly Englishman* (London: Trübner & Co.), p. 305.

195 **'Antilopes [sic] are set':** Ibid.

196 **As in hawking, one bird:** Captain Thomas Williamson, *Oriental Field Sports* (Naval & Military Press).

199 **Daniel Johnson:** Daniel Johnson, in 1827, wrote one of the first books on field sports in India.

199 **hunting method of the cheetah:** Daniel Johnson, *Sketches of Indian Field Sports: with observations on the animals; also an account of some of the customs of the inhabitants; with a description of the art of catching serpents, as practised by the conjoors and their method of curing themselves when bitten: with remarks on hydrophobia and rabid animals* (London: Robert Jennings), p. 110.

199 **We reached a position:** Sir Samuel White Baker, *Wild Beasts and their Ways: Reminiscences of Europe, Asia, Africa and America* (London, New York: Macmillan and Co. Ltd.), p. 175.

201 **Sudugee, 19 November:** Walter Campbell, *My Indian Journal* (Edinburgh: Edmonston & Douglas), p. 78.

202 **A little after mid-day:** W. H. Sleeman, *Rambles and Recollections of an Indian Official* 1 (London: Archibald Constable and Company), p. 148.

202 **In my morning's ride:** W. H. Sleeman, *Rambles and Recollections of an Indian Official* 2 (London: Archibald Constable and Company), p. 9.

204 **The trumpeting noises:** Thomas Skinner, *Excursions in India: including a walk over the Himalaya Mountains, to the sources of the Jumna and the Ganges* 1 (London: Henry Colburn and Richard Bentley), p. 50.

204 **In the suburbs of the city:** Edward Delaval Hungerford Elers Napier, *Scenes and Sports in Foreign Lands* (London: Henry Colburn), p. 174.

205 **I saw a whole street:** Marianne North, *Recollections of a Happy Life: Being the Autobiography of Marianne North*, ed. John Symonds (New York: Macmillan and Co. Ltd.).

205 **flighty nature:** Richard Holmes, *Sahib: The British Soldier in India 1750-1914* (London: HarperCollins), p. 165. He gives an example of the Indian cheetah's tame nature: 'Albert Hervey's regiment the officers were often accompanied by sepoys… Indeed, his adjutant owed his life to a sepoy's intervention. A wounded cheetah had seized the officer by his cap…when the sepoy ran up and killed the creature with a mighty crack across the head with his hunting knife.'

205 **the maharaja's palace in Mysore:** Henry Bevan, *Thirty Years in India: or, A soldier's reminiscences of native and European life in the presidencies, from 1808 to 1838* 1 (P. Richardson), p. 220. The rajah of Mysore and the Hon. A. H. C resident, both of whom contributed most liberally to promote the public amusements at Bangalore, used to send several cheetahs to Mysore during the races, the hunting of which afforded good sport on the course. Spearing them off horseback was the favourite mode of attack, but the skin of this animal is so loose and tough that it was difficult to penetrate—this is why severe scratches and other damage were inflicted on the animals before they could be killed.

205 **'The cheetah or hunting leopard is':** G. P. Sanderson, *Thirteen Years among the Wild Beasts of India* (London: W. H. Allen & Co.), p. 320.

205 **In fact, Sanderson did manage:** 'I also shot two cheetahs and a bear during the period of my command at Rhowra. One of these cheetahs was shot close to my tent…' Henry Bevan, *Thirty Years in India: or, A soldier's reminiscences of native and European life in the presidencies, from 1808 to 1838* 1 (P. Richardson), p. 126.

206 **One day, shooting:** Ibid., p. 159.

206 **'It struck me as rather odd':** William Rice, *Tiger-shooting in India: being an account of hunting experiences on foot in Rajpootana, during the hot seasons, from 1850 to 1854* (London: Smith, Elder and Co.).

207 **Sometime after my return:** Gordon-Cumming, William Gordon, *Wild Men and Wild Beasts* (New York: Scribner, Armstrong), p. 335.

208 **On one occasion:** William Rice, *"Indian Game,": (from Quail to Tiger)* (London: W. H. Allen & Co.), p. 78.

208 **I once speared:** David Newall, *The Highlands of India* 2 (A. Brannon and Son), p. 448.

210 **I was out walking:** Osborn Wilkinson and Johnson Wilkinson, *The Memoirs of the Gemini Generals: Personal Anecdotes, Sporting Adventures, and Sketches of Distinguished Officers* (London: A. D. Innes), p. 102.

211 **Among the beasts:** Sir Edward Braddon, *Thirty Years of Shikar* (W. Blackwood & Sons).

212 **At Jeraikela:** A. Mervyn Smith, *Sport and Adventure in the Indian Jungle* (London: Hurst and Blackett), p. 279.

212 **Servants lead tame; One of the tame cheetahs:** Pierre Loti, *India* (New Delhi: Asian Educational Services, first published in London), pp. 192, 199.

212 **It is generally believed:** A. Mervyn Smith, *Sport and Adventure in the Indian Jungle* (London: Hurst and Blackett Publishers).

213 **R. G. Burton confirms:** R. G. Burton, *The Tiger Hunters* (London: Hutchinson & Co.). The hunting leopard lives mainly on antelope, and the red lynx or caracal preys on smaller animals and birds; but these two, both of which in captivity are used in the chase, are seldom found by the sportsman, for they are rare animals.

213 **The Hunting Leopard:** The Indian Empire, *The Imperial Gazetteer of India: Descriptive* 1 (Oxford: Clarendon Press).

214 **recounts how cheetahs:** Major Henry Shakespear, *The Wild Sports of India* (London: Smith, Elder and Co.), p. 202. 'The coursing of antelope with the hunting leopard is a pretty sport, and is much followed by the wealthy natives of India: it used to be followed by European officers also. The price of a well-broken hunting cheetah is from one hundred and fifty to two hundred and fifty rupees.'

214 **'was sighted'; 'Some people say that':** Divyabhanusinh, *The End of a Trail: The Cheetah in India* (New Delhi: Oxford University Press), p. 146.

214 **According to Divyabhanusinh:** Ibid, p. 155.

215 **They were being bred:** 'Cheetah-Racing—It's a New Sport: The Indian Cheetah is the World's Fastest Animal', *The Argus Week-end Magazine* (London, 10 January 1938).

215 **...the Baroda manual:** Divyabhanusinh, *The End of a Trail: The Cheetah in India* (New Delhi: Oxford University Press), p. 137.

215 **Further, if an animal:** Ibid., p. 124.

215 **The possibility:** Ibid., p. 123.

215 **It can be argued:** Ibid., p. 137.

216 **Cat Specialist Group in 1984:** The valuable piece of information about non-existence of the hunting cheetah in Jaipur has been discovered by Chuttan Khan and Chotte Khan, both claiming to be young lads of the Cheetah Training School of Jaipur at the end of the last century. Authentic breeding records of the Jaipur zoo strongly supplement the stray field observations of the hunters and the authors.

216 **Existence of the cheetah:** The authors in communication with the Maharaja of Baroda, Shri Fateh Singhji Gaekwad.

217 **Hopes for reintroduction:** V. Sharma and K. Sankhala, 'Vanishing Cats of Rajasthan', Proceedings from the Cat Specialist Group meeting in Kanha National Park, ed. P. Jackson, 1984, pp. 116-135.

218 **At the turn of the twentieth:** Frank Finn, *Sterndale's Mammalia of India* (Bombay: Thacker & Co.).

218 **so had Pocock:** R. I. Pocock, *The Fauna of British India, Including Ceylon and Burma* (London: Taylor and Francis Ltd.).

218 **'made a living':** Rowland Ward, *Records of Big Game: containing an account of their distribution, descriptions of species, lengths, and weights, measurements of horns and field notes* (London: Rowland Ward and Co. Limited).

218 **Hook...divested himself:** John Pinfold, 'Cheetah Racing Remembered: An Exploration Through the Sources', *African Research & Documentation* no. 111 (Standing Conference on Library Materials on Africa, 1 January 2009).

220 **'Meantime his study':** John Pollard, *African Zoo Man—The Life Story of Raymond Hook* (London: Robert Hale), p. 99.

220 **'For purposes of':** A. Mervyn Smith, *Sport and Adventure in the Indian Jungle* (London: Hurst and Blackett), p. 279.

220 **It is commonly supposed:** Major General Nigel Woodyatt, *My Sporting Memories: Forty years with Note-book & Gun* (H. Jenkins).

221 **I was guiding a raja:** J. A. Hunter, *Hunter* (London: Hamish Hamilton), p. 55.

221 **These fabulously wealthy men:** Ibid., p. 209.

222 **'His Highness sat':** C. J. P. Ionides, *A Hunter's Story: Autobiography* (London: Allen).

222 **The maharaja was:** Alastair Scobie, *Animal Heaven* (Cassell & Co.), p. 123.

223 **Not only big game:** Ibid.

224 **What follows therefore:** Anthony Cullen and Sydney Downey, *Saving the Game* (London: Jarrolds Publishers Ltd.), p. 133.

224 **'were so nauseated':** Bombay Natural History Society, *Journal of the Bombay Natural History Society* 47 (Bombay: Bombay Natural History Society, 1948), p. 719.

BIBLIOGRAPHY

'Al Khanzir', Lieut-Gen., H. G. Martin. *Sunset from the Main.* Suffolk: Richard Clay and Company Ltd., 1951.

Adams, Andrew. *Wanderings of a Naturalist in India: The Western Himalayas and Cashmere.* Edinburgh: Edmonston and Douglas, 1867.

Adamson, George. *My Pride and Joy*. Glasgow: Collins Harvill, 1986.

Alexander, R. and Martin-Leake, A. *Some Signposts to Shikar.* Calcutta: Fred. R. Grenyer, 1932.

Ali, M. Athar. *The Mughal Nobility under Aurangzeb*. Delhi: Oxford University Press, 1996.

Allen, Charles and Dwivedi, Sharada. *Lives of the Indian Princes.* Mumbai: Eshwar, 1998.

Allen, Hugh. *The Lonely Tiger.* London: Faber and Faber, 1960.

Allsen, Thomas T. *The Royal Hunt in Eurasian History*. Philadelphia: University of Pennsylvania Press, 2006.

Alvi, M. A. and Rahman, A. *Jahangir the Naturalist.* New Delhi: The National Institute of Sciences of India, 1968.

Alvi, M. A. and Rahman. A. *Jehangir: The Naturalist.* New Delhi: National Institute of Science of India, 1968.

Anderson, J. K. *Hunting in the Ancient World.* Berkeley: University of California Press, 1985.

Anderson, Kenneth. *Jungles Long Ago.* London, New Delhi: Rupa Publications India Pvt. Ltd. in arrangement with George Allen and Unwin Ltd., 1976.

Anderson, Kenneth. *Man-Eaters and Jungle Killers.* London, New Delhi: Rupa Publications India Pvt. Ltd. in arrangement with George Allen and Unwin Ltd., 1957.

Anderson, Kenneth. *The Black Panther of Sivanipalli.* London, New Delhi: Rupa Publications India Pvt. Ltd. in arrangement with George Allen and Unwin Ltd., 1957.

Anderson, Kenneth. *The Tiger Roars.* London, New Delhi: Rupa Publications India Pvt. Ltd. in arrangement with George Allen and Unwin Ltd., 1967.

Anderson, Kenneth. *This is the Jungle.* London, New Delhi: Rupa Publications India Pvt. Ltd. in arrangement with George Allen and Unwin Ltd., 1976.

Anonymous i.e. Bakewell Robert. *Stray Leaves from the Diary of an Indian Officer.* London: Whitfield, Green and Son, 1865.

Archer, Edward Caulfield. *Tours in Upper India, and in Parts of the Himalaya Mountains, With Accounts of the Courts of the Native Princes.* London: Richard Bentley, 1833.

Archer, Mildred. *Company Paintings: Indian Paintings of the British Period.* London, Ahmedabad: Simultaneously published by Victoria and Albert Museum in association with Mapin Publishing Pvt. Ltd., 1992.

Armstead, Christopher. *Princely Pageant.* Thomas Harmsworth Publishing, 1987.

Atkins, E. H. *A Naturalist on the Prowl.* London and Calcutta: W. Thacker and Co., 1897.

Atkins, E. H. *The Tribes on My Frontier.* Calcutta: Thacker, Spink and Co., 1914.

Atkinson, G. F. *Curry and Rice: British Social Life in 19th Century.* New Delhi: Time Books International, 1982.

Babur. *Baburnama.* Fifteenth century.

Bacon, Thomas. *First Impressions and Studies from Nature in Hindostan.* London: W. H. Allen and Co., 1837.

Baden-Powell, Sir Robert. *Memories of India.* London: H. Jenkins, 1915.

Bagot, A. *Sport and Travel in India and Central* America. London: Chapman & Co., 1897.

Baillie, Mrs W. W. *Days and Nights of Shikar.* London: John Lane, 1921.

Baker, E. C. Stuart. *Mishi the Man-Eater.* London: H. F. & G. Witherby Ltd., 1928.

Baker, S.W. *The Rifle and the Hound in Cashmere.* New York: Arno Press, 1850.

Baker, Sir Samuel W. *Wild Beasts and their Ways.* London: Macmillan and Co., 1898.

Ball, Valentine. *Jungle Life in India; or, the Journeys and Journals of an Indian Geologist.* London: Thos. De La Rue and Co., 1880.

Ball, Valentine. *Travels in India by Jean Baptiste Tavernier Baron of Aubonne.* London: Oxford University Press, 1925.

Banerjee, J. N. *The Development of Hindu Iconography.* Calcutta: University of Calcutta, 1956.

Banerji, Sures Chandra. *Flora and Fauna in Sanskrit Literature.* Calcutta: Naya Prokash, 1980.

Barclay, Edgar N. *Big Game Shooting Records.* London: H. F. & G. Witherby, 1932.

Barnum, Phineas T. *Jack in the Jungle.* Amsterdam: Fredonia Books, 1908.

Barnum, Phineas T. *The King of the Animal Kingdom.* Chicago: R. S. Peale & Co., 1889.

Barras, Colonel Julius. *India and Tiger-Hunting.* London: Swan Sonnenschein & Co., 1885.

Barton, S.W. *The Princes of India and Nepal.* Jaipur: Pointer Publishers, 2007.

Basham, A. L. *The Wonder that was India: A Survey of History and Culture of the Indian Sub-Continent Before the Coming of the Muslims.* London: Sidgwick & Jackson, 1982.

Baskaran, S. Theodore. *The Dance of the Sarus.* New Delhi: Oxford University Press, 1999.

Basu, Dr Helene. *Cultural Practices of Indian Sidis through the Prism of Indian Ocean Maritime Connections.* Bombay: For private circulation, 2006.

Bayly, C.A. *Rulers, Townsmen and Bazaars.* New Delhi. Oxford University Press .

Beach, Milo C. Fischer, Eberhard. Goswamy B. N. *Masters of Indian Painting 1100-1650.* Volume I. Germany: Artibus Asiae Publishers, 2011.

Beach, Milo C. Fischer, Eberhard. Goswamy B. N. *Masters of Indian Painting 1100-1650.* Volume II. Germany: Artibus Asiae Publishers, 2011.

Beach, Milo Cleveland (ed.). *King of the World: The Padshahnama: an Imperial Mughal*

Manuscript from the Royal Library, Windsor Castle. Azimuth Ed., 1997.

Beal, Samuel (trans.). *Travels of Fah-Hian and Sung-Yun*. London: Trübner & Co., 1869.

Beard, Peter. *The End of the Game*. Germany: Taschen Hohenzollernring, 2008.

Bedi, Ramesh. *Corbett National Park*. New Delhi: Clarion Books, 1985.

Berinstain, Valerie. *Mughal India: Splendours of the Peacock Throne*. London: Thames and Hudson, 1998.

Bernier, Francois. *Travels in the Mogul Empire: 1656-1668*. London: W. Pickering, 1826.

Best, James W. *Forest Life in India*. London: John Murray, 1935.

Best, James W. *Indian Shikar Notes*. Lahore: The Pioneer Press, 1920.

Best, James W. *Tiger Days*. London: John Murray, 1931.

Betzig, Laura. *Hunting Kings*. University of Michigan; Sage Publications, 2008.

Bevan, Henry. *Thirty Years in India: or, a Soldier's Reminiscences of Native and European Life in the Presidencies, from 1808 to 1838*. London: Pelham Richardson, 1839.

Big Game Hunting in India and the Game Animals of India. Manager of Publications, 1948.

Birdwood, Marshall. *Khaki and Gown*. London: Ward, Lock & Co., 1941.

Blaffer, Sarah. *The Langurs of Abu*. Harvard University Press, 1977.

Blake, Stephen P. *Shahjahanabad: The Sovereign City in Mughal India, 1639-1739*. Cambridge: Cambridge University Press, 1991.

Blane, William, Somerville, William and Squire, Francis. *Cynegetica; or, Essays on sporting: consisting of observations on hare hunting: ...Together with an account of the Vizier's manner of hunting in the Mogul Empire*. London: Printed for John Stockdale, 1788.

Blane, William. *An Account of the Hunting Excursions of Asoph ul Doulah: Nabob of Oudh*. London: Printed for John Stockdale, 1788.

Blaze, William. *Tiger! Tiger!* Britain: Elek Books Limited, 1957.

Bloch, J. *Les Inscriptions d'Asoka*. Paris: Les Belles Lettres, 1950.

Blood, Sir Bindon. *Four Score Years and Ten*. London: G. Bell and Sons, 1933.

Bluchel, Kurt G. *Game and Hunting*. Volume II. Germany: Konemann Verlagsgesellschaft mbH, 1997.

Blyth, E. *Journal of the Asiatic Society of Bengal: Mammals and Birds of Burma in II Parts*. Hertford: Stephen Austin and Sons, 1875.

Blyth, Edward. *Catalogue of the Mammalia in the Museum*. Calcutta: Asiatic Society, published by order of the Society, 1863.

Boddington, Craig. *From Mt. Kenya to the Cape*. Long Beach: Safari Press, 1987.

Bombay Natural History Society, *Journal of the Bombay Natural History Society*. Volume 47. Bombay: Bombay Natural History Society, 1948.

Booth, Martin. *Carpet Sahib: A Life of Jim Corbett*. London: Constable, 1986.

Bore, Big. *Guide to Shikar on the Nilgiris*. Madras: S. P. C. K. Depot, 1924.

Bostock, John Ashton. *Letters from India and the Crimea etc*. London: George Bell and Sons, 1896.

Braddon, Edward Nicholas Coventry. *Life in India: A Series of Sketches Showing Something of the Anglo-Indian, etc*. London: Longmans, Green & Co., 1872.

Braddon, Sir Edward. *Thirty Years of Shikar*. London: William Blackwood and Sons, 1895.

Bradley, Mary Hastings. *Trailing the Tiger.* London: D. Appleton and Company, 1929.

Brander, A. A. Dunbar. *Wild Animals in Central India.* London: Edward Arnold & Co., 1927.

Breasted, J. H. *A History of Egypt.* London: Hodder & Stoughton Limited, 1906.

Brinckman, Arthur. *The Rifle in Cashmere: a narrative of shooting expeditions in Ladak, Cashmere, Punjaub, etc., with advice on travelling, shooting, and stalking: to which are added notes on army reform and Indian politics.* London: Smith, Elder and Co., 1862.

Brown, J. Moray. *Shikar Sketches with Notes on Indian Field Sports.* London: Hurst and Blackett Publishers, 1887.

Brown, J. Moray. *Stray Sport.* London: William Blackwood and Sons, 1893.

Brown, Samuel Sneade. *Home Letters Written from India.* London: C. F. Roworth, 1878.

Brown, W. Llewellyn. *The Etruscan Lion.* Oxford: Clarendon Press, 1960.

Browne, E. G. *A Year Amongst the Persians: Impressions as to the Life, Character, & Thought of the People of Persia.* Cambridge: Cambridge University Press, 1926.

Buchanan, Francis *A Journey from Madras through the Countries of Mysore, Canara and Malabar.* London: W. Bulmer and Co., 1807.

Buck, Frank and Lehrer, Steven (ed.). *Bring 'em Back Alive.* USA: Texas Tech University Press, 2000.

Buck, Frank. *On Jungle Trails.* London: George G. Harrap & Co. Ltd., 1938.

Bull, Bartle. *Safari: A Chronicle of Adventure.* New York: Carroll & Graf Publishers, reprinted in 2006.

Bull, H. M. and Haksar, K. N. *Madhav Rao Scindia of Gwalior: 1876-1925.* Gwalior: Alijah Darbar Press, 1926.

Burger, John F. *African Adventures.* London: Robert Hale Limited, 1957.

Burghall, George. *Letter from Major Burghall to the Honourable The Directors of the East India Company.* 1778.

Burke, Norah. *Jungle Picture.* Bombay: Allied Publishers, 1973.

Burke, W. S. *The Indian Field Shikar Book.* Calcutta: Thacker Spink and Co., 1920.

Burnell Arthur Coke and Tiele, P. A. (ed.). *The Voyage to the East Indies.* Volume I. London: Hakluyt Publications, 1985.

Burt, Thomas Seymour. *Narrative of a Late Steam Voyage from England to India via the Mediterranean. (Part II: Account of a Late Palankeen Trip from Bombay to Mhow and Lahore.)* Calcutta: 1840.

Burton, Brig-Gen R. G. *A Book of Man-Eaters.* New Delhi: Mittal Publications, 1984.

Burton, Brig-Gen R. G. *Sport and Wildlife in the Deccan.* London: Seeley, Service and Co. Limited, 1928.

Burton, Brig-Gen R. G. *The Book of the Tiger.* London: Hutchinson and Co. Publishers Ltd., 1933.

Burton, Brig-Gen R. G. *The Tiger Hunters.* London: Hutchinson and Co. Publishers Ltd, 1936.

Burton, Edmund. *An Indian Olio.* London: Spencer Blackett, 1888.

Butler, John. *A Sketch of Assam.* London: Smith, Elder and Co., 1847.

Butt, Khan Saheb Jamshed. *Shikar.* London: Robert Hale Limited. 1963.

Cameron Joseph F. S. MacDowall. *Chutney Papers: Society, Shikar, and Sport in India.*

Bombay: W. Thacker and Co., 1884.

Campbell, Joseph. *The Hero with a Thousand Faces*. New York: Pantheon Books, 1949.

Campbell, Major Walter. *The Old Forest Ranger*. London: George Routledge and Sons, Ltd., 1845.

Campbell, Walter. *My Indian Journal*. Edinburgh: Edmonston and Douglas, 1864.

Canby, Sheila R. *Persian Masters: Five Centuries of Painting*. Mumbai: Marg Publications, 1990.

Carnac, J. H. Rivett. *Many Memories of Life in India, at Home and Abroad*. London: William Blackwood and Sons, 1910.

Casserly, Gordon. *Life in an Indian Outpost*. London: T. W. Laurie, 1914(?).

Cavaliere, Barbara. *Wonder of the Age: Master Painters of India 1100-1900*. New York: Metropolitan Museum of Art, 2011.

Chakra, Suhaash. *The Raj Syndrome*. New Delhi: Rupa Publications India Pvt. Ltd., 2007.

Chakrabarti, Kalyan. *Man-Eating Tigers*. Calcutta: Darbari Prokashan, 1992.

Champion, F. W. *The Jungle in Sunlight and Shadow*. London: Chatto & Windus, 1933.

Champion, F. W. *With a Camera in Tiger-Land*. London: Chatto & Windus, 1927.

Channel, A. R. *Jungle Rescue*. London: Dennis Dobson, 1967.

Chaudhuri, K. N. *Jheel and Jungle*. Calcutta: Thacker, Spink and Co., 1918.

Choudhury, Suklambari. *Khairi: The Beloved Tigress*. Dehradun: Natraj Publishers, 1999.

Chubb, William John Jeremy. *The Lucknow Menagerie: Natural History Drawings from the Collection of Claude Martin (1735-1800)*.

Colonel Burton. *Tigers of the Raj*. Gloucester: Alan Sutton, 1987.

Commissariat, M. S. *Mandelslos Travels in Western India (1638-1639)*. London: Oxford University Press, 1931.

Coomaraswamy, A. K. *History of Indonesian and Indian Art*. New York: 1927.

Copland, Ian. *The Princes of India in the Endgame of Empire 1917-1947*. New Delhi: Cambridge University Press India Ltd., 1999.

Corbett, Jim. *Jungle Lore*. London: Oxford University Press, 1953.

Corbett, Jim. *Man-Eaters of Kumaon*. New Delhi: Oxford University Press, 1944.

Corbett, Jim. *My India*. New Delhi: Oxford India Paperbacks, 1952.

Corbett, Jim. *The Man-Eating Leopard of Rudraprayag*. New Delhi: Oxford India Paperbacks, 1947.

Corbett, Jim. *The Mohan Man-Eater and Other Stories*. London: Oxford University Press, 1948.

Corbett, Jim. *The Temple Tiger and More Man-Eaters of Kumaon*. New Delhi: Oxford India Paperbacks, 1954.

Coryate, Thomas. *Coryat's Crudities: reprinted from the edition of 1611*. London: W. Cater, 1776.

Coryate, Thomas. *Coryatt's Crudities*. Volume VIII. Reprinted from the 1611 edition.

Craighead, John J. & Craighead, Frank C. Jr. *Life with an Indian Prince*. Idaho: Archives of American Falconry, 2001.

Crill, Rosemary and Jariwala, Kapil. *The Indian Portrait 1560-1860*. Ahmedabad: Mapin Publishing, 2010.

Crill, Rosemary. *Marwar Painting: A History of Jodhpur Style*. Mumbai: India Book House, 2001.

Crooke, William and Tod, James. *Annals and Antiquities of Rajasthan*. Volume I and II. London: Oxford University Press, 1920.

Cullen, Anthony and Downey, Sydney. *Saving the Game*. London: Jarrolds Publishers Limited, 1960.

Curtis, John. *The Oxus Treasure (Objects in Focus)*. London: British Museum Press, 2012.

Curzon, Lord. *A Viceroy's India*. London: Sidgwick & Jackson, 1984.

Dalrymple, William. *The Last Mughal: The Fall of a Dynasty*. Delhi: Penguin Books, 2006.

Daniel, J. C. *A Century of Natural History*. Bombay: Bombay Natural History Society, 1983.

Daniel, J. C. *The Leopard in India*. Dehradun: Natraj Publishers, 1996.

Daniel, J. C. *The Tiger in India: A Natural History*. Dehradun: Natraj Publishers, 2001.

Daniell, Thomas. *Oriental scenery: 24 Views of Hindoostan Taken in the Year 1792*. London: 1792.

Daniell, William and Caunter, Hobart. The Oriental Annual, or Scenes in India. London: Bull and Churton, 1836.

Danvers, Robert William. *Letters from India to China 1854-58*. London: Hazell, Watson, and Viney, Ltd.

Das, Giridhari Prasad. *India-West Asia: Trade in Ancient Times—6th Century BC to 3rd Century AD*. New Century Publications, 2006.

Dass, Diwan Jarmani. *Maharaja: The Lives, Loves and Intrigues of the Maharajas of India*. New Delhi: Hind Pocket Books Pvt. Ltd., 2007.

Davidar, E. R. C. *Cheetal Walk*. New Delhi: Oxford University Press, 1997.

Davids, Rhys T. W. and C. A. F. (trans.). *Dialogues of the Buddha*. Volume III. London: The Pali Text Society, 1921.

Davidson, Charles James C. *Diary of Travels and Adventures in Upper India, from Bareilly, in Rohilcund, to Hurdwar, and Nahun, in the Himmalaya Mountains, with a Tour in Bundelcund, a Sporting Excursion in the Kingdom of Oude, and a Voyage Down the Ganges*. London: Henry Colburn, 1843.

Davidson, Lieut-Col David. *Memories of a Long Life*. Edinburgh: David Douglas, 1890.

de Kloguen, Denis. *A Historical Sketch of Goa*. Madras: Printed for the proprietor by William Twigg at the Gazette Press, 1831.

de Romanis, F. and Tchernia A. (eds.). *Crossings: Early Mediterranean Contacts with India*. New Delhi: Manohar, 1997.

Dehejia, Vidya with contributors. *India Through the Lens: Photography 1840-1911*. Freer Gallery of Art and Arthur M. Sackler Gallery (Smithsonian Institution), Washington DC, in association with Mapin Publishing and Mandala Publishing, 2000.

Delort, Robert. *The Life and Lore of the Elephant*. London: Thames and Hudson, 1992.

Department of Environment. *Indira Gandhi on Environment*. New Delhi: Govt. of India, 2006.

Deraniyagala, P. E. P. *Some Extinct Elephants, Their Relatives and the Two Living Species*. Ceylon: Ceylon National Museums Publication, 1955.

Devahuti, D. *The Unknown Hsuan Tsang*. New Delhi: Oxford University Press, 2001.

Devee, Maharanee Sunity C. I. *Bengal Dacoits and Tigers*. Calcutta: Thacker, Spink and Co., 1916.

Dewar, Douglas. *Birds of the Indian Hills*. London: John Lane, 1915.

Dewar, Douglas. *Indian Birds' Nests*. Calcutta: Thacker, Spink and Co., 1929.

Dinerstein, Eric. *Tigerland and Other Unintended Destinations*. Washington: Island Press, 2005; originally published in 1970.

Diver, Maud. *Royal India*. New York: D. Appleton & Company, 1942.

Divyabhanusinh. *The End of a Trail: The Cheetah in India*. New Delhi: Banyan Books, 1995.

Divyabhanusinh. *The Lions of India*. New Delhi: Black Kite, an imprint of Permanent Black, 2008.

Divyabhanusinh. *The Story of Asia's Lion*. Mumbai: Marg Publications, 2005.

Doshi, Saryu. *India and Egypt*. New Delhi: Marg Publications.

Doyle, C. W. *The Taming of the Jungle*. Westminster: Archibald Constable and Co., 1899.

Dunbar, Janet. *Tigers, Durbars and Kings: Fanny Eden's Indian Journals*. London: John Murray, 1988.

Durand, Sir Edward Law. *Rifle, Rod and Spear in the East*. London: John Murray, 1911.

Eardley-Wilmot, S. *The Life of a Tiger*. London: Edward Arnold & Co., 1911.

Edward, J. Drew Gay. *The Prince of Wales in India; or, from Pall Mall to the Punjab*. Toronto: Belford Brothers, 1877.

Edwardes, Michael. *British India 1772-1947*. New Delhi: Rupa Publications India Pvt. Ltd., 1967.

Edwardes, Michael. *Glorious Sahibs*. London: Eyre & Spottiswoode, 1968.

Edwards, I. E. S., Gadd, C. J. and Hammond, N. G. *Cambridge Ancient History*. Volume I. Cambridge: Cambridge University Press, 1970.

Edwards, I. E. S., Solleberger, E. and Hammond, N. G. *Cambridge Ancient History*. Volume II. Cambridge: Cambridge University Press, 1973.

Egerton, Francis. *Journal of a Winter's Tour in India: With a Visit to the Court of Nepaul*. Volume I and II. London: John Murray, 1852.

Elliot, Robert H. *Gold, Sport and Coffee Planting in Mysore*. Westminster: Dodo Press, 1898.

Elliott, Major-General J. G. *Field Sports in India 1800-1947*. London: Gentry Books Limited, 1973.

Ellis, Edward S. *The Jungle Fugitives*. New York: Hurst and Company, 1903.

Ellison, Bernard C. *H. R. H. The Prince of Wales's Sport in India*. London: William Heinemann, Ltd., 1925.

Elwood, Anne Katherine. *Narrative of a journey overland from England, by the continent of Europe, Egypt and the Red sea to India, including a residence there and voyage home*. Volume 2. London: Henry Colburn and Richard Bentley, 1830.

Epigraphia Indica. Volume I. Archaeological Survey of India, 1882.

Eraly, Abraham. *The Mughal World: India's Tainted Paradise*. London: Phoenix, 2008.

Ernst, Waltraud and Pati, Biswamoy. *India's Princely States: People, Princes and Colonialism*. London: Routledge, 2007.

Etherton, Colonel P.T. *The Last Stronghold*. London: Jarrolds, 1934.

Evans, George Patrick Elystan. *Big Game Shooting in Upper Burma*. London: Longmans, Green, and Co., 1910.

Fane, Henry Edward. *Five Years in India*. London: Henry Colburn, 1842.

Fayrer, J. *The Royal Tiger of Bengal: His Life and Death*. London: J. & A Churchill, 1875.

Fazl, Abu'l. *Ain-i-Akbari*. Sixteenth century CE.

Fazl, Abu'l. *Akbarnama*. Sixteenth century CE.

Fenton, Lt. Col. L. L. *The Rifle in India*. London: W. Thacker and Co., 1923.

Festing, Gabrielle. *Strangers within the Gates*. Edinburgh: William Blackwood and Sons, 1914.

Field, D. M. *Jungle Jottings*. Jodhpur: Jodhpur Government Press.

Fife-Cookson, Lt. Col. J. C. *Tiger-shooting in the Doon and Ulwar with Life in India*. London: Chapman and Hall, 1887.

Finn, Frank. *Sterndale's Mammalia of India*. Calcutta: Thacker, Spink and Co., 1929.

Finn, Frank. *The Birds of Calcutta*. Calcutta: Thacker, Spink and Co., 1917.

Fisher, Michael H. *Visions of Mughal India: An Anthology of European Travel Writing*. London: I. B. Tauris, 2007.

Fitz, Sir Kenneth Samuel. *Twilight of the Maharajas*. London: John Murray, 1956.

Fitzroy, Yvonne. *Courts and Camps in India*. London: Methuen and Co. Ltd., 1926.

Fletcher, F. W. F. *Sport on the Nilgiris and in Wynaad*. London. Macmillan and Co. Ltd., 1911.

Foran, Major W. Robert. *Kill: or be Killed*. London: Hutchinson and Co. Publishers Ltd., 1933.

Forbes, Gordon Sullivan. *Wild Life in Canara and Ganjam*. London: Swan Sonnenschein & Co., 1885.

Forbes, James and Montalambert, Eliza. *Oriental Memoirs by J. Forbes, revised by his daughter Countess de Montalambert*. London: Richard Bentley, 1834.

Forsyth, Captain J. *The Highlands of Central India*. London: Chapman and Hall, 1919.

Foster, Sir William. *Early Travels in India, 1583-1619*. Oxford: Oxford University Press, 1921.

Foster, William. *The Journal of John Jourdain 1608-1617*. Cambridge: Printed for the Hakluyt Society, 1905.

Foster, William. *The Voyage of Thomas Best to the East Indies 1612-14*. London: Printed for the Hakluyt Society, 1934.

Fraser, James Baillie. *Journal of a Tour*. New Delhi: Rupa Publications India Pvt. Ltd., 1814.

Fraser, Sir Andrew. *Among Indian Rajahs and Ryots*. London: Seeley and Co. Ltd., 1911.

Fraser, Thomas and Malleson, George. *Records of Sport and Military Life in Western India*. London: W. H. Allen and Co., 1881.

Fraser, W. M. *The Recollections of a Tea Planter*. London: The Tea and Rubber Mail, 1935.

Frayer, Sir Joseph. *Recollections of My Life*. London: William Blackwood and Sons, 1900.

Frazer, Sir James. *The Golden Bough: A Study in Magic and Religion.* Volume I. New York: Macmillan, 1894.

Fryer, John. *A New Account of East-India and Persia in Eight Letters.* Volume I, II, III. London: Printed for the Hakluyt Society, 1909.

G. Aflalo F. *A Book of Wilderness and Jungle with Big Game Hunting Anecdotes.* London: S.W. Partridge & Co., Ltd., 1912.

Gardner, Nora Beatrice Blyth. *Rifle and Spear with the Rajpoots: Being the Narrative of a Winter's Travel and Sport.* London: Chatto & Windus, 1895.

Garga, D. P. *From My Big Game Diaries.* Calcutta: The Book Company Ltd., 1944.

Gayatri Devi. *A Princess Remembers.* New Delhi: Rupa Publications India Pvt. Ltd, 1995.

Geddie, John. *Beyond the Himalayas.* London: T. Nelson and Sons, 1884.

Gerard, Jules. *Lion Hunting: Adventures and Exploits of Famous Hunters and Travellers.* Ward, Lock and Co., 1856.

Gerard, Lt. Gen. Sir Montagu Gilbert. *Leaves from the Diaries of a Soldier and Sportsman during 20 Years Service in India, Afghanistan, Egypt and Other Countries, 1865-1885.* London: John Murray, 1903.

Gershevitch, Ilya. *Cambridge History of Iran.* Volume II. Cambridge University Press, 1985.

Gibb. Sir H. *The Travels of Ibn Batuta AD 1325-1354.* Volume III. London: Cambridge University Press, 1971.

Gilhus, Ingvild Saelid. *Animals, Gods and Human: Changing Attitudes to Animals in Greek, Roman and Early Christian Ideas.* England: Routledge, 2006.

Gillmore, Parker. *Leaves from a Sportsman's Diary.* London: Gibbons and Company Limited, 1896.

Gilmour, David. *The Ruling Caste: Imperial Lives in the Victorian Raj.* London: John Murray, 2005

Glasfurd, A. I. R. *Leaves from an Indian Jungle.* Bombay: The Times Press, 1903.

Glasfurd, A. I. R. *Musings of an Old Shikari.* London: John Lane, 1928.

Glasfurd, A. I. R. *Rifle and Romance in the Indian Jungle.* London: John Lane, 1905.

Gordon, Charles Alexander. *Through Three Victorian Campaigns: Experiences of a Regimental Surgeon during the Gwalior War, Campaigns in West Africa and the Indian Mutiny.* Leonaur, 2009.

Gordon, Iain. *Soldiers of the Raj: The Life of Richard Purvis 1789-1868.* South Yorkshire: Leo Cooper, an imprint of Pen and Sword, 2001.

Gordon-Cumming, William Gordon. *Wild Men and Wild Beasts.* London: Scribner, Armstrong, 1872.

Gouldsbury, C. E. *Tigerland.* New York: E. P. Dutton & Co., 1916.

Graham, B. N. Gordon. *Hunter at Heart.* London: Herbert Jenkins, 1950.

Graham, G. *Life in the Mofussil: or, the Civilian in Lower Bengal.* London: C. Kegan Paul & Co., 1878.

Gray, Albert assisted by H. C. P. Bell. *The Voyage of Francois Pyrard of Laval to the East Indies, the Maldives, the Moluccas and Brazil in theree parts.* London: Printed for the Hakluyt Society, translated from the French edition of 1619.

Gray, James. *Life in Bombay, and the Neighbouring Outstations etc.* London: Richard Bentley, 1852.

Graydon, William Murray. *The Jungle Boy.* Maryland: Wildside Press, 2007.

Greenwood, James. *Wild Sports of the World.* London: S. O. Beeton, 1862.

Grey, C. and Garrett H. L. O (ed.). *European Adventurers of Northern India 1785-1849.* Sussex: The Naval and Military Press, reprint 2009.

Grey, Edward. *The Travels of Pietro Della Valle in India: From the Old English Translation of 1664 by G. Havers.* London: Printed for the Hakluyt Society, 1892.

Grove, R. H. *Green Imperialism: Colonial Expansion, Tropical Island Edens and the Origins of Environmentalism 1600-1860.* Cambridge: Cambridge University Press, 1995.

Grube, Ernst J. *A Mirror for Princes from India.* Mumbai: Marg Publications, 1991.

Guggisberg, C. A. W. *Early Wildlife Photographers.* Oxford: David and Charles, 1800.

Guggisberg, C. A. W. *Simba: The Life of the Lion.* South Africa: Chilton Books, 1961.

Guggisberg, C. *Simba, the Life of the Lion.* Cape Town: Howard Timmins, 1961.

Gunn, Hugh M. A. *Empire Big Game.* London: Simpkin, Marshall, Hamilton, Kent and Co. Ltd., 1925.

Habib, Irfan. *An Atlas of the Mughal Empire: Political and Economic Maps with Detailed Notes, Bibliography and Index.* New Delhi: Oxford University Press, 1982.

Hahn, Daniel. *The Tower Menagerie.* London: Pocket Books, 1937.

Haidar, Navina Najat and Sardar, Marika. *Sultans of the South: Arts of India's Deccan Courts, 1323-1687.* New York: Metropolitan Museum of Art, 2011.

Hall, Basil. *Travels in India, Ceylon and Borneo.* London: George Routledge and Sons, Ltd., 1931.

Hamilton, Alexander. *A New Account of the East Indies 1688-1723.* Volume I, II. London: Printed for A. Bettesworth and C. Hitch, 1727.

Handley, Leonard M. H. *Hunter's Moon.* London: Macmillan and Co. Ltd., 1933.

Hanley, Patrick. *Tiger Trails in Assam.* London: Robert Hale Limited, 1961.

Harris, Captain W. Cornwallis. *Portraits of the Game and Wild Animals of South Africa.* Alberton: Galgo Publishing, 1986.

Harrison, Eric. *Gunners, Game and Gardens.* London: Leo Cooper, 1978.

Harry R. Caldwell. *Blue Tiger.* New York: Duckworth, 1924.

Hasan, A. K. M. *In the Jaws of a Tiger.* Calcutta: W. Newman & Company, 1971.

Hasan, Farhat. *State and Locality in Mughal India: Power Relations in Western India.* Cambridge: Cambridge University Press, 2004.

Haughton, H. L. *Sport and Folklore in the Himalaya.* Edinburgh: Ballantine, Hanson & Co., 1913.

Hay, Sidney. *Historic Lucknow.* Lucknow: The Pioneer Press, 1939.

Hazra, R. C. *Studies in the Upapuranas.* Volume II. Calcutta: 1958.

Heather, Red. *Memories of Sporting Days.* London: Longmans, Green & Co., 1923.

Heber, Bishop, Laird, M. A. (ed.). *Bishop Heber in Northern India.* New York: Cambridge University Press, 1971.

Heber, Reginald. *Narratives of a Journey Through the Upper Provinces of India: From Calcutta to Bombay, 1824-1825, (with Notes upon Ceylon,) an Account of a Journey to Madras and the Southern Provinces, 1826, and Letters Written in India.* London: John

Murray, 1828.

Heck, Lutz. *Animals My Adventure*. London: Methuen & Co Ltd., 1954.

Hewett, Sir John. *Jungle Trails in Northern India*. London: Methuen and Co. Ltd., 1938.

HH The Maharaja of Mysore. *African Survey*. The Bangalore Press.

Hilton, Richard. *Nine Lives*. London: Hollis & Carter, 1955.

Hinton, Anthony (ed. and trans.). *Wild Animals of Africa*. London: Neville Spearman, 1956.

Hodges, William. *Travels in India during the years 1780-83*. London: Printed for the author, sold by J. Edwards, 1783.

Hodges-Hill, Edward. *Man Eater: Tales of Lion and Tiger Encounters*. Heathfield: Cockbird Press, 1992.

Holmes, Richard. *Sahib: The British Soldier in India*. London: HarperCollins, 2005.

Hooker, Joseph Dalton. *Notes of a tour in the Plains of India, the Himalaya and Borneo*. London: Reeve, Benham, and Reeve, 1848.

Hornaday, William Temple. *Two Years in the Jungle: The Experiences of a Hunter and Naturalist in India, Ceylon, the Malay Peninsula and Borneo*. New York: Charles Scribner's Sons, 1885.

Horton, Barbara Curtis. *Tiger Bridge*. Santa Barbara: John Daniel & Co., 1993.

Hoyland, J. S. and Banerjee, S. N. *The Commentary of Father Monserrate*. New Delhi: Asian Educational Services, reprinted in 2003.

Hudson, Rev. George H. *Twelve Years of a Soldier's Life in India*. Boston: Ticknor & Fields, 1860.

Hughes, Albert William. *Outlines of Indian History*. 1861.

Hughes, Julie Elaine. *Animal Kingdoms: Princely Power, the Environment and the Hunt in Colonial India*. Harvard University Press, 2013.

Hunter, J. A. *Hunter*. London: Hamish Hamilton, 1952.

Hunter, J. A. *Hunter's Tracks*. London: Hamish Hamilton, 1957.

Hunter, Sir William Wilson. *The Indian Empire, its People, History, and Products*. London: Trübner & Co., 1886.

Hutchisson, W. H. Florio and Wilson, John. *Pen and Pencil Sketches: Being Reminiscences during Eighteen Years' Residence in Bengal*. London: Sampson Low, Marston, Searle, & Rivington, 1883.

Imam, S. A. *Brown Hunter*. Allied Publishers, 1979.

Inglis, James. *Sport and Work on the Nepaul Frontier*. Westminster: Dodo Press, 1878.

Inglis, James. *Tent Life in Tigerland and Sport and Work on the Nepaul Frontier*. London. Sampson Low, Marston, Searle, & Rivington, 1892.

Inglis, James. *Tent Life in Tigerland*. London: Sampson Low, Marston, Searle, & Rivington, 1888.

Ionides, C. J. P. *A Hunter's Story*. London: W. H. Allen and Co., 1965.

Ions, Veronica. *Egyptian Mythology*. London: The Hamlyn Publishing Group Ltd., 1965.

Islam, Shuja Ul and Roush, John H. Jr. (ed.). *Hunting Dangerous Game with the Maharajas*. New Delhi: Himalayan Books, 2000.

Islam, Zohra and Islam, Shuja Ul. *Hunting Dangerous Game with the Maharajas in the Indian Sub-Continent*. New Delhi: Himalayan Books, 2004.

Ismail, Lt. Col. M. M. *Call of the Tiger*. London: Faber and Faber, 1964.
Ives, Edward. *A Voyage from England to India, in the Year MDCCLIV; or, A Journey from Persia to England by an Unusual Route*. London: Printed for Edward and Charles Dilly, 1754.
Ives, Richard. *Of Tigers and Men: Entering the Age of Extinction*. New Delhi: Nan A. Talese, 1996.
Jackson, Deirdre. *Lion*. London: Reaktion Books, 2010.
Jacquemont, Victor. *Letters from India describing a journey in the British dominions of India, Tibet, Lahore, and Cashmere, during the years 1828, 1829, 1830, 1831, undertaken by order of the French government*. Volume II. London: Edward Churton, 1834.
Jain, S. K., Agarwal, P. K. and Singh, V. P. *Hydrology and Water Resources in India*. Netherlands: Springer, 2007.
Jardine, William on Baron Cuvier's work. *Lions, Tigers, & c., & c.* London: H. G. Bohn, 1846.
Jeffery, Roger and Nandini Sundar. *A New Moral Economy for India's Forests*. New Delhi: Sage Publications, 1999.
Jennison, George. *Animals for Show and Pleasure in Ancient Rome*. USA: Manchester University Press, 1937.
Jepson, Stanley. *Big Game Encounters*. London: H. F. & G. Witherby Ltd., 1936.
Jepson, Stanley. *More Big Game Encounters*. Volume II. Bombay: W. Thacker and Co., 1944.
Johnsingh, A. J. T. *Field Days: A Naturalist's Journey through South and Southeast Asia*. Hyderabad: Universities Press India Pvt. Ltd., 2005.
Johnsingh, A. J. T. *On Jim Corbett's Trail*. New Delhi: Permanent Black, 2004.
Johnson, Daniel. *Sketches of Indian Field Sports: with observations on the animals; also an account of some of the customs of the inhabitants; with a description of the art of catching serpents, as practised by the conjoors and their method of curing themselves when bitten: with remarks on hydrophobia and rabid animals*. London: Robert Jennings, 1827.
Johnstone, J. W. D. *The Gwalior of Scindia's*. Spring Books, 2004.
Joseph, J. D. *Life in the Wilds of Central India*. Edinburgh: Empire and Commonwealth Museum, 1930.
Joslin, Paul. *The Asiatic Lion: A Study of Ecology and Behaviour*. University of Edinburgh, 1973.
Jung, Colonel Sher. *Tryst with Tigers*. New Delhi: Hind Pocket Books, in arrangement with Robert Hale.
Jung, Saad Bin. *Wild Tales from the Wild*. New Delhi, London: Lotus Collection, Roli Books India Pvt. Ltd., 2005.
Jung, Sher. *Tryst with Tigers*. New Delhi, London: Hind Pocket Books, in arrangement with Robert Hale Limited.
Kalof, Linda. *Looking at Animals in Human History*. London: Reaktion Books, 2007.
Kangle, R. P. *The Kautilya Arthasastra*. Mumbai: University of Bombay, 1965.
Kanjilal, U. N. and Kanjilal, P. C. *Flora of Assam*. The Govt. of India, 1934.
Karageorghos, V. (ed.). *The Greeks Beyond the Aegean: From Marseilles to Bactria*. New York: Alexander S. Onassis Foundation, 2002.

Karanth, K. Ullas. *A View from the Machan*. New Delhi: Permanent Black, 2006.

Karanth, K. Ullas. *The Way of the Tiger*. Hyderabad: Universities Press India Pvt. Ltd., 2001.

Karanth, K. Ullas. *Tiger Tales*. New Delhi: Penguin India, 2006.

Kaye, Sir John William. *Lives of Indian Officers: Illustrative of the History of the Civil and Military Service of India*. Volume I, II. London: A. Strahan & Co. & Bell and Daldy, 1867.

Kearton, Cherry. *In the Land of the Lion*. London: J. W. Arrowsmith Ltd., 1943.

Kearton, Cherry. *Photographing Wildlife across the World*. London: J. W. Arrowsmith Ltd., 1913.

Kennedy, Hugh. *When Baghdad Ruled the Muslim World*. Cambridge: Da Capo Press, 2005.

Kennion, Major R. L. *By Mountain, Lake and Plains*. London: William Blackwood and Sons, 1911.

Khan, Mutamad. *Tuzuk-i-Jahangiri*.

Khan, Sahibzada Abdul Shakoor. *Shikar Events and Some Useful Notes Thereon*. New Delhi: 1935.

Khan, Sahibzada Abdul Shakoor. *Wild Life and Hunting*. New Delhi: Light and Life Publishers, 1978.

Khanam, Zaheda. *Birds and Animals in Mughal Miniature Paintings*. New Delhi: DK Printworld, 2009.

Kincaid, C. A. *Forty-four Years a Public Servant*. London: William Blackwood and Sons, 1934.

Kincaid, Dennis. *British Social Life in India 1608-1937*. London: George Routledge and Sons Ltd., 1938.

Kingston, R. W. G. *A Naturalist in Himalaya*. London: H. F. & G. Witherby Ltd., 1920.

Kingston, W. H. G. *Adventures in India*. London: George Routledge and Sons, Ltd., 1884.

Kingston, W. H. G. *The Young Rajah*. London: T. Nelson and Sons, 1876.

Kipling, John Lockwood. *Beast and Man in India*. London: Macmillan and Co. Ltd., 1891.

Kipling, Rudyard. *The Second Jungle Book*. London: Macmillan & Co. Ltd., 1957.

Knighton, William. *The Private Life of an Eastern King*. New York: Redfield, 1855.

Knowles, George Hogan. *In the Grip of the Jungles*. London: Wright and Brown, 1932.

Kossak, Steven. *Indian Court Painting: 16th-19th Century*. London: Thames and Hudson, 1997.

Krishnan, M. and Guha, Ramachandra (ed.). *Nature's Spokesman: M. Krishnan and the Indian Wildlife*. New Delhi: Penguin Books, 2007.

Kumar, K. G. Pramod. *Posing for Posterity: Royal Indian Portraits*. New Delhi: Roli Books India Pvt. Ltd., 2012.

Kyle, Donald G. *Sport and Spectacle in the Ancient World*. London: Blackwell Publishing, 2007.

Lady Lawrence. *Indian Embers*. Palo Alto: Trackless Sands Press, 1991.

Lahiri-Choudhury, Dhriti K. *The Great Indian Elephant Book*. New Delhi: Oxford University Press, 1999.

Laisram, Pallavi Pandit. *Viewing the Islamic Orient: British Travel Writers of the Nineteenth Century.* London: Routledge, 2006.

Lal, Muni. *Jahangir.* New Delhi: Vikas Publishing House Pvt. Ltd., 1983.

Langley, Edward Archer. *Narrative of a Residence at the Court of Meer Ali Moorad; with Wild Sports in the Valley of the Indus.* London: Hurst and Blackett Publishers, 1860.

Lanz, Tobias J. *On the Trail of the Indian Tiger.* USA: Safari Press, 2009.

Larking, Col. Cuthbert. *Bandobast and Khabar.* London: Hurst and Blackett Publishers, 1888.

Le Messurier, Col. A. *Game, Shore and Water Birds of India.* London: W. Thacker and Co., 1904.

Leatham, A. E. *Sport in Five Continents.* London: William Blackwood and Sons, 1912.

Lee-Warner, William. *The Protected Princes of India.* London, New York: Macmillan and Co. Ltd., 1894.

Legge, James. *A Record of Buddhistic Kingdoms: Being an Account by the Chinese Monk Fa-Hien of his travels in India and Ceylon (A. D. 399-414) in Search of the Buddhist Books of Discipline; Translated and Annotated with a Corean Recension of the Chinese Text.*

Leidy, Denise Patry with contributions by Sherman E. Lee. *Treasures of Asian Art: The Asia Society's Mr. and Mrs. John Rockefeller 3rd Collection.* New York: The Asia Society Galleries, 1994.

Leveson, H. A. *Sport in Many Lands.* London: Chapman and Hall, 1877.

Linschoten, Jan Hugyen Van. *Voyage to Goa and Back, 1583-1592.* Volume I, II. Westminster: Archibald Constable and Co Ltd., 1903.

Linschoten, Jan Huygen Van, Burnell, A. C. and Tiele, Pieter Anton. *The voyage of John Huyghen van Linschoten to the East Indies: from the old English translation of 1598: the first book, containing his description of the East.* London: Printed for the Hakluyt Society, 1885.

Liu, Xinru. *The Silk Road in World History.* USA: Oxford University Press, 2010.

Locke, J. Courtenay. *The First Englishmen in India: Letters and Narratives of Sundry Elizabethans Written by Themselves.* 1920.

Locke, Lt. Col. A. *The Tigers of Trengganu.* London: Museum Books Limited, 1954.

Lockwood, Edward. *Natural History, Sport and Travel.* London: W. H. Allen and Co., 1878.

Losty, J. P. and Roy, Malini. *Mughal India: Art, Culture and Empire.* London: British Library, 2012

Losty, J. P., Khurshid, Salman, Nanda, Ratish and Singh, Malvika. *Delhi: Red Fort to Raisina.* New Delhi: Roli Books India Pvt. Ltd., 2012.

Loti, Pierre and Inman, George A. (trans.). *India.* New Delhi: Asian Educational Services, 1995 (reprint).

Lt. Rice. *Tiger-Shooting in India.* London: Smith, Elder and Co., 1857.

Lydekker, Richard. *Lloyd's Natural History: A Handbook to the Carnivora.* London: Edward Lloyd, 1896.

Lydekker, Richard. *The Great and Small Game of India, Burma and Tibet.* London: Rowland Ward, 1900.

Mackay, R. D. *Have You Shot an Indian Tiger?* New Delhi: Sikkson's Press, 1967.

MacPherson, Donald. *The Raj: A Time Remembered*. Durham: Pentland Press, 2000.

Mainwaring, Henry Germain. *A Soldier's Shikar Trips*. London: Grant Richards Ltd., 1920.

Malcolm, Sir John. *A Memoir of Central India*. London: Printed for Kingsbury, Parbury, & Allen, 1824.

Mannin, Ethel. *Jungle Journey*. London: Jarrolds Publishers Limited, 1948.

Manucci, Niccolao and Irvine (trans.). *A Pepys of Mogul India 1653-1708: An Abridged Edition of Storia Do Mogor*. New York: E. P. Dutton, 1913.

Manucci, Niccolao and Irvine (trans.). *Storia do Mogor: or, Mogul India (1653-1708)*. London: John Murray, 1907.

Marozzi, J. *Tamerlane: Sword of Islam, Conqueror of the World*. London: HarperCollins, 2004.

Marshall, Edison. *Shikar and Safari*. London: Museum Press Ltd., 1930.

Martin, D. King. *The Ways of Man and Beast in India*. London: Wright and Brown, 1935.

Martin, Monica. *Out in the Midday Sun*. Boston: Little, Brown and Co., 1949 (reprint).

Martinelli, Antonio and Michell, George. *Oriental Scenery: Two Hundred Years of India's Artistic and Architectural Heritage*. London: Swan Hill Press, an imprint of Airlife Publishing Limited, 1998.

Masters, Brian. *Maharana: The Story of the Rulers of Udaipur.* Ahmedabad: Mapin Publishing, 1990.

Matyushkin, E. (ed.). *The Amur Tiger in Russia: An Annotated Bibliography*. Moscow: World Wildlife Fund, 1998.

May, Sir Edward S. *Changes & Chances of a Soldier's Life*. London: Philip Allan & Co., 1925.

Maydon, Major H. C. *Big Game of India*. Volume XXIII. London: Philip Allan & Co., 1937.

McCrindle, J. W. *Ancient India as Described in Classical Literature*. Amsterdam, 1971.

McCrindle, J. W. *The Invasion of India by Alexander the Great*. London: Kessinger Publishing, reprinted from 1916.

McLaughin, Raoul. *Rome and the Distant East*. London: Continuum, 2010.

Miller, Lt. Col. E. D. *Fifty Years of Sport*. New York: E. P. Dutton and Co., 1925 (?).

Ministry of Transport. *With Gun and Rod in India*. New Delhi: The Tourist Traffic Branch, Ministry of Transport.

Mirza, H. and Phillott, L. C. (trans.). *Baznama i-Nasiri: A Persian Treatise on Falconry*. London: Bernard Quaritch, 1908.

Mitchell, K. W. S. *Tales from some Eastern Jungles*. C. Palmer, 1928.

Mitchell, Mrs Murray. *In India: Sketches of Indian Life and Travel from Letters and Journals*. Nelson, 1876.

Mitra, Sudipta. *Gir Forest and the Saga of the Asiatic Lion*. New Delhi: Indus Publishing Company, 2005.

Mitra, Sudipta. *History and Heritage of Indian Game Hunting*. New Delhi: Rupa Publications India Pvt. Ltd., 2010.

Mitter, P. *Indian Art*. Oxford: Oxford University Press, 2001.

Montgomery, S.Y. *Spell of the Tiger.* Boston: Houghton Mifflin Company, 1995.

Mookerji, Radhakumud. *A History of Indian Shipping.* 1912.

Moosvi, Shireen. *People, Taxation and Trade in Mughal India.* New Delhi: Oxford University Press, 2008.

Morgan, John Hartman. *Leaves from a Field Notebook.* London: Macmillan and Co. Ltd., 1916.

Morrow, Ann. *The Maharajas of India.* New Delhi: Srishti Publishers and Distributors, 1998.

Mrs Fenton by Henry Lawrence. *The Journal of Mrs Fenton.* New York: Cambridge University Press, 1820.

Mundy, Captain. *The Journal of a Tour in India.* Volume I, II. London: John Murray, 1832.

Munro, Innes. *A Narrative of the Military Operations, On the Coromandel Coast.* London: Printed for the author by T. Bensley, 1789.

Munshi, K. M. *The End of an Era (Hyderabad Memoirs).* Bombay: Bharatiya Vidya Bhavan, 1957.

Musselwhite, Arthur. *Behind the Lens in Tigerland.* Calcutta: Thacker, Spink and Co., 1933.

Naidu, Kamal. *Trail of the Tiger.* Dehradun: Oxford University Press, 1998.

Napier, Edward Hungerford Delaval Elers. *Scenes and Sports in Foreign Lands.* London: Henry Colburn, 1840.

Nawabs of India, including: Nawab Sadeq Mohammad Khan V, Azam Jah, Osman Ali Khan, Asaf Jah VII. Wajid Ali Shah, Nawab of Bhopal, Muhammad Jalaluddin Khan, Amin Ud-Din Ahmad Khan, Alivardi Khan. Nawab of the Carnatic, Khwaja Salimullah, Chandra Sahib. Hephaestus Books, 2011.

Neill, John. *Recollections of Four Years service in the East, with H. M. Fortieth Regiment.* London: Richard Bentley, 1845.

Neumayer, Erwin. *Lines on Stone: The Prehistoric Rock Art of India.* New Delhi: Manohar Publishers and Distributors, 1993.

Newall, D. *The Highlands of India.* Volume I, II. 1882.

Newall, J. T. *Scottish Moors and Indian Jungles.* London: Hurst and Blackett Publishers, 1889.

Newall, J.T. *The Eastern Hunters.* London: Tinsley brothers, 1886.

Nicholls, Frank. *Assam Shikari.* New Zealand: Tonson Publishing House, 1971.

Nichols, Andrew. *Ctesias: On India.* London: Bristol Classical Press, 2011.

Nizami, Khaliq Ahmad. *Royalty in Medieval India.* Delhi: Munshiram Manoharlal Publishers Pvt. Ltd., 1997.

North West Province, Late Customs' Officer. *Past Days in India: Or, Sporting Reminiscences of the Valley of the Soane and the basin of Singrowlee.* London: Chapman and Hall, 1874.

North, Marianne and Symonds, John (ed.). *Recollections of a Happy Life: Being the Autobiography of Marianne North.* New York: Macmillan and Co. Ltd., 1894.

Nott, J. F. *Wild Animals: Photographed and Described.* London: Sampson Low, Marston, Searle & Rivington, 1886.

O'Brien, Stephen J. *Tears of the Cheetah: And Other Tales from the Genetic Frontier.* New York: Thomas Dunne Books, an imprint of St Martin's Press, 2003.
O'Dwyer, Sir Michael. *India as I knew it.* London: Constable & Company Ltd., 1925.
O'Meara, Lt. Col. E. J. *I'd live it Again.* J. B. Lippincott Co., 1898.
O'Shea, F. Bernard. *A Winter's Tour in India and Ceylon with a Kathiawar Prince.* Bombay: Printed at the "Times of India" Steam Press, 1890.
Oaten, E. F. *European Travellers in India.* London: Kegan Paul, Trench, Trübner & Co. Ltd., 1909.
Officer in India. *Letter from an officer in India to his correspondent in England.* London. Printed for J. Debrett, 1794.
Ojha, Dr Geeta. *Economic History of India: Trade during the Great Mughals 1526-1707 AD.* New Delhi: Shri Sai Printographers, 2005.
Oldfield, Henry Ambrose. *Sketches from Nepal.* Volume I, II. London: W. H. Allen and Co., 1880.
Oliver, Daniel. *First Book of Indian Botany.* London: Macmillan and Co., 1869.
Oppenheim, A. L. *Ancient Mesopotamia: Portrait of a Dead Civilization.* Chicago: University of Chicago Press, 1964.
Osmaston, B. B. *Wild Life and Adventures in Indian Forests: from diaries.* Henry Osmaston, 1977.
Outram, James. *A Memoir of the Public Services Rendered by Lieutenant Colonel Outram.* London: Smith, Elder and Co., 1853.
Outram, James. *Baroda Intrigues and Bombay Khutput.* London: Smith, Elder and Co., 1853.
Outram, James. *Rough Notes of the Campaign in Sinde and Afghanistan: In 1838-39.* Bombay: American Mission Press, 1840.
Outram, James. *The Conquest of Scinde.* Edinburgh: William Blackwood and Sons, 1846.
Ovington, J. and Guha, J. P. (ed.) *India in the Seventeenth Century.* New Delhi: Associated Publishing House, 1984.
Ovington, J. and Rawlinson, H. G. (ed.). *A Voyage to Surat in the Year 1689.* London: Oxford University Press, 1696.
Paget, Leopold Grimston. *Camp and Cantonment: A Journal of Life in India in 1857-1859, With Some Account of the Way Thither.* London: Longman, Green, Longman, Roberts & Green, 1865.
Pal, Pratapaditya. *Court Paintings of India 16th-19th Centuries.* New York: Navin Kumar, 1983.
Pal, Pratapaditya. *Painted Poems: Rajput Paintings from the Ramesh and Urmil Kapoor Collection.* Mumbai: Mapin Publishing, 2004.
Palit, Major-General D. K. *Musings and Memories.* Volume I, II. New Delhi: Palit and Palit in association with Lancer Publishers, 2004.
Parks, Fanny. *Wanderings of a Pilgrim.* Volume I, II. London: Pelham Richardson, 1850.
Patnaik, Naveen. *A Second Paradise: Indian Courtly Life 1590 -1947.* London: Sidgwick & Jackson Ltd., 1985.
Paul, E. Jaiwant and Kapoor, Pramod. *One Hundred and Fifty Years of Photography: The*

Unforgettable Maharajas. New Delhi: Roli Books India Pvt. Ltd., 2012.

Peissel, Michel. *Tiger for Breakfast*. Bombay: Allied Publishers, 1966.

Pennant, Thomas. *The Views of Hindoostan*. Volumes I-IV. London: Printed by Henry Hughs, 1798.

Percy, Lt. Col. Reginald Heber. *Big Game Shooting in India: The Early Days*. Read Books.

Perry, Richard. *The World of the Tiger*. London: Cassell, 1964.

Pester, John. *War and Sport in India, 1802-06: An Officer's Diary*. London: Heath, Cranton and Ouseley Ltd.

Petrie, Flinders W. M. *The Making of Egypt*. London: Sheldon Press, 1939.

Pliny, the Elder by John Healy. *Natural History: A Selection*. London: Penguin Classics, 1991.

Pocock, R. I. *The Fauna of British India, including Ceylon and Burma*. London: Taylor and Francis Ltd., 1939.

Pollard, John. *Africa for Adventure*. London: Robert Hale Limited, 1961.

Pollard, John. *African Zoo Man*. London: Robert Hale Limited, 1963.

Pollock, Lt. Col. William. *Sport in British Burmah, Assam, and the Cassyah and Jyntiah Hills*. London: Chapman and Hall, 1879.

Pollock, Lt. Col. William. *Sporting Days in Southern India*. London: Horace Cox, 1894.

Pollok, Lt. Col. William. *Incidents of Foreign Sport and Travel*. London: Chapman and Hall, 1894.

Pollok, Lt. Col. William. *Wild Sports of Burma and Assam*. London: Hurst and Blackett Publishers, 1900.

Porter, J. Hampden. *Wild Beasts*. New York: Charles Scribner's Sons, 1894.

Porter, John Hampden. *Wild Beasts*. Charles Scribner's Sons, 1891.

Potter, William Simpson. *Letters from India during H.R.H. the Prince of Wales' Visit in 1875-76 from WS Potter to his Sister*. London: E. Lawless, 1876.

Powell, A. N. W. *Call of the Tiger*. London: Robert Hale Limited, 1957.

Prater, S. H. *The Book of Indian Animals*. Bombay: Bombay Natural History Society, 1948.

Proske, Roman. *My Turn Next: The Autobiography of an Animal Trainer*. London: Museum Press Ltd., 1957.

Pulley, Louis Henry. *Five Years in India: Being an Account of Arthur Herbert in the Land of "Caste and Curries"*. London: C. S. Burbidge, 1857.

Quammen, David. *Monster of God*. London: W. W. Norton & Company, 2003.

R. S. Dharma Kumar Sinhji of Bhavnagar. *Reminiscences of Indian Wildlife*. New Delhi: Oxford University Press, 1998.

Rajyashree Kumari, Bikaner. *The Maharajas of Bikaner*. New Delhi: Amarylis, 2012.

Rangarajan, Mahesh. *India's Wildlife History*. New Delhi: Permanent Black in association with The Ranthambhore Foundation, 2001.

Rangarajan, Mahesh. *The Oxford Anthology of Indian Wildlife* Volume I, II. New Delhi: Oxford University Press, 1999.

Raoul. *Small Game Shooting in Bengal*. Calcutta: W. Newman & Co., 1899.

Rashid, M. A. and David, Reuben. *The Asiatic Lion*. Baroda: Under a MAB Project financed by Department of Environment, Govt. of India, 1992.

Rathore, L. S. Professor. *The Regal Patriot: Maharaja Ganga Singh of Bikaner.* New Delhi: Lotus Collection, Roli Books India Pvt. Ltd., 2007.

Reade, Julian. *Assyrian Sculpture.* London: British Museum Press, 1998.

Reddy, E. Ajaikumar. *Man-eating Tigers of Central India.* New Delhi: Indialog Publications, 2004.

Reid, Mayne. *Plant Hunters.* New York: John W. Lowell Company, 1869.

Ricci, Aldo. *The Travels of Marco Polo.* New Delhi: Rupa Publications India Pvt. Ltd., 1931.

Rice, Major-General William. "*Indian Game,": (from Quail to Tiger).* London: W. H. Allen and Co., 1884.

Robbins, Kenneth and McLeod, John (ed.). *African Elites in India: Habshi Amarat.* Ahmadabad: Mapin, 2006.

Robinson, Andrew. *Maharaja.* London: Thames & Hudson, 1988.

Robinson, Bradley. *World's Great Stories of Hunting and Adventure.* London: W. H. Allen and Co., 1947.

Roe, Sir Thomas and Fryer, Dr John. *Travels in India in the Seventeenth Century: reprinted from the Calcutta Weekly Englishman.* New Delhi: Asian Educational Services, 1993 (reprint).

Ross, Mark C. and Reesor, David. *Predator: Life and Death in the African Bush.* New York: Abrams, an imprint of Harry N. Abrams, 2007.

Rothfeld, Otto. *With Pen and Rifle in Kashmir.* New Delhi: Reprinted by Anmol Publications Pvt. Ltd., 1993.

Rothfels, Nigel. *Savages and Beasts: The Birth of the Modern Zoo.* London: Johns Hopkins University Press, 2002.

Rousselet, Louis. *India and Its Native Princes: Travels in Central India and in the Presidencies of Bombay and Bengal.* New Delhi: Asian Educational Services, 1875.

Roy, Jit. *Shikar Tales by the Barrel.* Bombay: Pearl Books, 1967.

Russell, C. E. M. *Bullet and Shot in Indian Forest, Plain and Hill.* London: W. Thacker and Co., 1900.

Rutley, C. Bernard. *Wild Life in the Jungle.* London: Macmillan & Co. Ltd., 1954.

Ryhiner, Peter as told to Mannix, Daniel P. *The Wildest Game.* Cassell, 1958.

Ryley, J. Horton. *Ralph Fitch: England's Pioneer to India and Burma: His Companions and Contemporaries.* London: T. Fisher Unwin, 1899.

Sainthill Eardley-Wilmot, C. I. E. *Forest Life and Sport in India.* London: Edward Arnold & Co., 1911.

Sanderson, G. P. *The Wild Beasts of India.* Edinburgh: John Grant, 1907.

Sanderson, G. P. *Thirteen Years among the Wild Beasts of India.* London: W. H. Allen and Co., 1896.

Sankhala, Kailash. *Return of the Tiger.* India: Lustre Press Pvt. Ltd., 1993.

Sankhala, Kailash. *Tiger! The Story of the Indian Tiger.* New Delhi: Rupa Publications India Pvt. Ltd. in arrangement with Collins, 1978.

Savory, Isabel. *A Sportswoman in India.* London: Hutchinson and Co. Publishers Ltd., 1900.

Savyasaachi. *Between the Earth and the Sky.* New Delhi: Penguin India, 2005.

Schafer, Edward H. *The Golden Peaches of Samarkand: A Study of T'ang Exotics*. Berkeley: University of California Press, 1963.

Schaller, George B. *The Deer and the Tiger*. Chicago: Chicago University Press, 1967.

Schaller, George B. *The Serengeti Lion*. Chicago: The University of Chicago Press, 1972.

Schimmel, Annemarie. *The Empire of the Great Mughals: History, Art and Culture*. London: Reaktion Books, 2004.

Schmitz, Barbara. *After the Great Mughals: Painting in Delhi and the Regional Courts in the 18th and 19th Centuries*. Mumbai: Marg Publications, 2002.

Scindia, Madhav Rao. *A Guide to Tiger Shooting*. 1920.

Scobie, Alastair. *Animal Heaven*. London: Cassell & Co Ltd., 1953.

Scott, Jack Denton. *Forests of the Night*. London: Robert Hale Limited, 1959.

Sen, S. N. *Travels of Jean De Thevenot and Gemelli Careri*. New Delhi: National Archives of India, Asian Educational Services reprint 2011.

Sergeant, Philip W. *The Ruler of Baroda*. London: John Murray, 1928.

Seshadri, Balakrishna. *Call of the Wild*. New Delhi: Sterling Publishers Limited, 1986.

Seshadri, Balakrishna. *The Twilight of India's Wildlife*. London: John Baker Publishers Limited, 1969.

Sewell, E. H. D. *An Outdoor Wallah*. London: Stanley Paul and Co. Ltd., 1945.

Shahi, S. P. *Backs to the Wall*. New Delhi: Affiliated East-West Press Pvt. Ltd., 1977.

Shakespear, Captain Henry. *The Wild Sports of India: With Remarks on the Breeding and Rearing of Horses and the Formation of Light Irregular Cavalry*. London: Smith, Elder & Co., 1860.

Shakespear, Major Henry. *The Wild Sports of India: With Detailed Instructions for the Sportsman; To Which Are Added Remarks on the Breeding and Rearing of Horses and the Formation of Light Irregular Cavalry*. London: Smith, Elder and Co., 1862.

Sharp, Sir Henry. *Good-Bye India*. New York: Oxford University Press, 1946.

Sheffield, Lt. Col. Frank. *How I Killed the Tiger*. London: Smith's Printing and Publishing Agency, 1902.

Sherer, J. W. *Daily Life During the Indian Mutiny*. Allahabad: Legend Publications, 1910.

Shivaramamurti, C. *The Art of India*. New York: Abrams, 1977.

Shukla, Rahul. *Killing Grounds: The Saga of Encounters in Wild*. New Delhi: Cosmo Publications, 1995.

Siddiqui, W. H. *Rampur Raza Library Monograph*. New Delhi: Rampur Raza Library, 1998.

Silver Hackle. *Indian Jungle Lore and the Rifle*. Calcutta: Thacker Spink and Co., 1929.

Silver, Hackle. *Man-Eaters and Other Denizens of the Indian Jungle*. Calcutta: Thacker Spink and Co., 1928.

Simson, Frank B. *Letters on Sport in Eastern Bengal*. London: R. H. Porter, 1886.

Sinclair, Gordon. *Foot-loose in India*. Toronto: S. B. Gundy, 1932.

Singh, Arjan. *A Tiger's Story*. New Delhi: HarperCollins Publishers India, 1999.

Singh, Arjan. *Tiger Haven*. London: Macmillan & Co. Ltd., 1973.

Singh, Colonel Kesri. *Hunting with Horse and Spear*. New Delhi: Hindustan Times Press, 1963.

Singh, Colonel Kesri. *One Man and a Thousand Tigers*. London: Robert Hale Limited, 1959.

Singh, Colonel Kesri. *One Man and Thousand Tigers*. New York: Dodd, Mead & Company, 1959.

Singh, Colonel Kesri. *The Tiger of Rajasthan*. Bombay: Rusi Khambatta R. M. D. C. Press, 1959.

Singh, Dr Ram Lakhan. *Tara: The Cocktail Tigress*. Allahabad: Print World, 1997.

Sinha, Vivek R. *The Tiger is a Gentleman*. Bangalore: Wildlife, 1999.

Sinha, Vivek R. *The Vanishing Tiger*. London: Salamander Books, 2003.

Skinner, James. *The Recollections of Skinner of Skinners Horse*. Leonaur, 2006.

Skinner, Thomas C. B. *Excursions in India: Including a Walk over the Himalaya Mountains to the Sources of the Jumna and the Ganges*. London: Henry Colburn and Richard Bentley, 1832.

Sleeman, Lt. Col. J. L. *Tales of a Shikari or Big Game Shooting in India*. London: Whitcombe and Tombs Limited, 1918.

Sleeman, Lt. Col. W. H. *Rambles and Recollections of an Indian Official*. Volume I, II. London: J. Hatchard and Son, 1844.

Sleeman, W. H. *A Journey through the Kingdom of Oude*. Volume I, II. London: Richard Bentley, 1859.

Smith, A. Mervyn. *Sport and Adventure in the Indian Jungle*. London: Hurst and Blackett Publishers, 1904.

Smith, T. Murray. *The Nature of the Beast*. London: Jarrolds Publishers Limited, 1963.

Smith, Vincent A. *The Early History of India: From 600 BC to the Muhammadan Conquest Including the Invasion of Alexander the Great*. Oxford: Clarendon Press, 1914.

Smuts, Field-Marshall J. C. *Jungle Man*. London: The Travel Book Club, 1947.

Smythies, E. A. *Big Game Shooting in Nepal*. Calcutta: Thacker, Spink & Co. Ltd., 1942.

Smythies, Olive. *Ten Thousand Miles on Elephants*. London: Seeley Service, 1961.

Smythies, Olive. *Tiger Lady*. Surrey: William Heinemann Ltd., 1953.

Snaffle (illustrated by Henry Dixon). *Gun, Rifle, and Hound in East and West*. London: Chapman and Hall, 1894.

Snilloc. *East African and Indian Jungle Thrillers*. Bombay: W. Thacker and Co., 1943.

Soho, Aristogeiton Marc. *Did the Lion Exist in Greece within Historic Times?* Johns Hopkins University, 1898.

Soifer, Deborah A. *The Myth of Narasimha and Vamana: Two Avatars in a Cosmological Perspective*. Albany: SUNY Press, 1991.

Somerville, Augustus. *Shikar near Calcutta*. Calcutta: W. Newman & Company, 1924.

Somerville, Augustus. *The Home of the Man-Eater*. London: Thacker, Spink & Co., 1933.

Somerville, Augustus. *The Passing of the Forest Gods*. Calcutta: Thacker, Spink and Co., 1933.

Spear, Percival. *The Nabobs: A Study of the Social Life of the English in Eighteenth Century India*. London: Oxford University Press, 1963.

Spodek, Howard. *Center for the Study of Federalism: Rulers, Merchants and Other Groups in the City-States of Saurashtra, India around 1800*. Philadelphia: Pennsylvania, Temple

University, reprinted from 1974.

Srivastav, Asheem and Suvira. *Asiatic Lion on the Brink*. Dehradun: Bishen Singh Mahendra Pal Singh, 1999.

Srivastava, Sanjeev. *Jahangir: A Connoisseur of Mughal Art*. New Delhi: Abhinav Publications, 2002.

Stacton, David. *A Ride on a Tiger: The Curious Travels of Victor Jacquemont*. London: Museum Press Ltd., 1954.

Stebbing, E. P. *Jungle By-Ways in India*. London: John Lane and The Bodley Head, 1911.

Stebbing, E. P. *The Diary of a Sportsman Naturalist*. London: John Lane, 1920.

Stephen, Oscar Leslie. *Sir Victor Brooke Sportsman and Naturalist: A Memoir of his Life and Extracts from his Letters and Journals.* London: John Murray, 1894.

Stephens, Martin. *Fair Game*. London: John Murray, 1925.

Sterndale, R. A. *Denizens of the Jungle: A Series of Sketches of Wild Animals, Illustrating their Forms and Natural Attitudes*. Calcutta: Thacker, Spink & Co., 1886.

Sterndale, Robert Armitage. *Natural History of the Mammalia of India and Ceylon*. New Delhi: Himalayan Books, 1884.

Sterndale, Robert Armitage. *Seonee; or, Camp life on the Satpura Range*. Calcutta: Thacker, Spink and Co., 1887.

Stewart, Colonel A. E. *Tiger and Other Game*. London: Longmans, Green & Co., 1928.

Stockley, Lt. Col. C. H. *Big Game Shooting in the Indian Empire*. London: Constable and Company Ltd., 1928.

Stone, Julia A. *Illustrated India: Its Princes and People.* Hartford, 1877.

Strabo by Jones, H. L. *The Geography of Strabo VII Books* XV-XVI. London, 1930.

Stracey, P. D. *Elephant Gold*. London: Weidenfeld and Nicolson, 1963.

Stracey, P. D. *The World of Animals: Tiger*. New York: Arthur Barker Limited, 1968.

Stracey, P. D. *Wildlife in India: Its Conservation and Control*. New Delhi: Govt. of India, 1963.

Strachan, Arthur W. *Mauled by a Tiger.* London: Moray Press, 1933.

Strong, J. S. *The Legend of King Asoka*. New Jersey: Princeton University Press, 1983.

Stronge, Susan. *Made for Mughal Emperors: Royal Treasures from Hindustan*. New Delhi: Roli Books India Pvt. Ltd. in arrangement with Roli & Janssen BV, 2010.

Stronge, Susan. *The Arts of the Sikh Kingdoms*. New Delhi: By V&A Publications for Prakash Book Depot, 1999.

Sucksdorff, Astrid Bergman. *Tiger in Sight*. London: Macmillan & Co. Ltd., 1965.

Sunquist, Fiona and Sunquist, Mel. *Tiger Moon*. Chicago: The University of Chicago Press, 2002.

Sunquist, Fiona and Sunquist, Mel. *Wild Cats of the World*. Chicago: University of Chicago Press, 2002.

Superintendent of Govt Printing. *Imperial Gazetteer of India: Rajputana and Central India*. Calcutta: 1908.

Sutherland, J. *Sketches of the Relations subsisting between the British Govt in India and the Different Native States.* Calcutta: G. H. Huttman, 1837.

Swaffham, John. 'Netting Tigers in the Jungle'. The Wide World Magazine. 1902.

Swarup, Shanti. *Flora and Fauna in Mughal Art.* India: D. B. Taraporevala Sons & Co. Pvt. Ltd., 1983.

Symington, John and Porterfield, Paul. *In a Bengal Jungle: Stories of Life on the Tea Gardens of Northern India.* North Carolina: The University of North Carolina Press, 1935.

Tavernier, Jean Baptiste. *Travels in India.* Volume I, II. London, New York: Macmillan & Co. Ltd., 1889.

Taylor, Major Neville. *Ibex Shooting on the Himalayas.* London: Sampson Low, Marston & Company, Ltd., 1903.

Taylor, Meadows. *The Story of my Life.* New Delhi: Asian Educational Services, reprinted from 1882.

Terry, Edward. *A voyage to East India wherein some things are taken notice of, in our passage thither, but many more in our abode there, within that ... empire of the Great Mogul.* London: Printed for W. Cater, reprinted from 1655.

Thapar, Romila. *Asoka and the Decline of the Mauryas.* New Delhi: Oxford University Press, 2012.

Thapar, Valmik. *Battling for Survival.* New Delhi: Oxford University Press, 2003.

Thapar, Valmik. *Tiger: Portrait of a Predator.* London: William Collins Sons and Co. Ltd., 1986.

Thapar, Valmik. *Tiger: The Ultimate Guide.* New York: CDS Books in association with Two Brothers Press, 2004.

The Government of India. *Big Game Hunting in India and the Game Animals of India.* New Delhi: India Press, 1948.

The Hon. Raja Kirtyanand Sinha. *Purnea: A Shikar Land.* Calcutta: Thacker, Spink and Co., 1916.

The Indian Empire. *The Imperial Gazetteer of India: Descriptive.* Volume I. Oxford: Clarendon Press, 1909.

The Maharaja of Cooch Behar. *Big Game Shooting in Eastern & North Eastern India.* New Delhi: Mittal Publications, 1985.

The Oxford India Illustrated Corbett. New Delhi: Oxford University Press, 2004.

The Wild Animals of the Indian Empire and the Problem of their Rewa, including: Tansen, Rajendra Nath, Shivanand, Rewa Prasad Dwivedi, Rewa, India, Rewa (princely state), Semaria, Deur Kothar, White Tiger, Hanumana, Rewa, Government Engineering College, Rewa, Awadhesh Pratap Singh University. Hephaestus Books, 1933.

Thomas, Henry Sullivan. *The Rod in India.* London: W. Thacker and Co., 1897.

Thomson, William. *Great Cats I Have Met: Adventures in Two Hemispheres.* Boston: Alpha Publishing Company, 1896.

Thornhill, Mark. *Haunts and Hobbies of an Indian Official.* London: John Murray, 1899.

Tod, James. *Annals and Antiquities of Rajasthan.* Volume III. Motilal Banarsidass, 1829.

Todd, W. Hogarth. *Tiger, Tiger!* London: Heath Cranton Limited, 1927.

Topsfield, Andrew. *Court Painting in Rajasthan.* Mumbai: Marg Publications, 2000.

Toynbee, Arnold. *A Study of History: The New One-Volume Edition.* London: Oxford University Press in association with Thames and Hudson Ltd., 1972.

Toynbee, J. M. C. *Animals in Roman Life and Art.* London: Thames and Hudson, 1973.

Tulloch, Maurice. *The All-in-One Shikar Book*. Bombay: D. B. Taraporevala Sons and Co. Ltd., 1950.

Turner, Alan, illustrations by Anton Mauricio. *The Big Cats and their Fossil Relatives.* New York: Columbia University Press, 1997.

Turner, Gurney. *First Impressions: or, A Day in India: A Letter from an Assistant Surgeon, Lately Arrived in Calcutta.* Yarmouth: Charles Sloman, 1841.

Turner, J. E. Carrington. *Man-Eaters and Memories.* Bombay: R. Khambatta Press reprinted from 1920.

Tyacke, Richard Humphry. *How I Shot My Bears; or, Two Years' Tent Life in Kullu and Lahoul*. London: Sampson Low, Marston, Searle, & Rivington, 1893.

Vaidya, Suresh. *Ahead Lies the Jungle*. London: Robert Hale Limited, 1958.

Vambery, A. *The Travels and Adventures of the Turkish Admiral Sidi Ali Reis: In India, Afghanistan, Central Asia and Persia during the years 1553-1556.* al-Biruni, 1975.

van Linschoten, Jan Huyghen, Burnell, Arthur Coke and Tiele, Pieter Anton. *The Voyage of John Huyghen van Linschoten to the East Indies: From the Old English Translation of 1598: The First Book, Containing His Description of the East.* London: Printed for the Hakluyt Society, 1885.

Van Linschoten, Jan Huyghen. *Voyage to Goa and Back.* Westminster: Archibald Constable and Co. Ltd., 1903.

Varaday, Desmond. *Gara-Yaka.* London: Collins, 1964.

Various. *Big Game Shooting in India around the Turn of the 20th Century.* Read Books Design, 2010.

Various. *Hunting and Shooting in India.* Read Books Design, 2000.

Various. *The Sportsman's Book for India.* London: Horace Marshall & Son, 1904.

Various. *Views of the Taje Mahel at the City of Agra, in Hindoostan.* BiblioBazaar, reprinted from 1789.

Verma, Som Prakash. *Flora and Fauna in Mughal Art.* Mumbai: Marg Publications, 1999.

Vigne, Godfrey Thomas. *Travels in Kashmir, Ladak, Iskardo: The Countries Adjoining the Mountain-Course of the Indus, and the Himalaya, North of the Panjab.* London: Henry Colburn, 1842.

Von Orlich, Cpt. Leopold. *Travels in India including Sinde and the Punjab.* Volume I, II. London: Longman, Brown, Green & Longmans, 1845.

Wakefield, William. *Our Life and Travels in India.* London: Sampson Low, Marston, Searle, & Rivington, 1878.

Wallace, Robert Grenville. *Fifteen Years in India; or Sketches of a Soldier's Life...* London: Longman, Hurst, Rees, Orme, and Brown, 1822.

Ward, Geoffrey C. and Ward, Diane Raines. *Tiger Wallahs.* New Delhi: Oxford University Press, 2002.

Ward, Rowland. *Records of Big Game: containing an account of their distribution, descriptions of species, lengths, and weights, measurements of horns and field notes.* London: Rowland Ward and Co. Limited, 1896.

Wardrop, Major-General A. E. *Days and Nights with Indian Big Game.* London: Macmillan and Co. Ltd., 1923.

Warmington, E. H. *The Commerce between the Roman Empire and India.* London:

Curzon Press; New York: Octagon Books, 1974.
Watters, Thomas. *On Yuan Chwang's Travels in India, 629-645 AD.* London: Royal Asiatic Society, 1904.
Weeden, Edward St Clair. *A Year with the Gaekwar of Baroda.* Boston: Dana Estes And Co., 1882.
Welch, Stuart Cary. *Gods, Kings and Tigers: The Art of Kotah.* Prestel, 1997.
Wheeler, James Talboys. *Early Records of British India.* London: Trübner & Co., 1878.
Wiki Series. *Maharajas of Rajasthan.* Memphis: 2011.
Wilkinson, Osborn & Wilkinson, Johnson. *The Memoirs of the Gemini Generals: Personal Anecdotes, Sporting Adventures.* London: A. D. Innes and Co., 1896.
Willcocks, General Sir James. *The Romance of Soldiering and Sport.* London: Cassell and Company, 1925.
Williams, J. H. *The Spotted Deer.* London: Rupert Hart-Davis, 1957.
Williams, J. *The Art of Gupta India.* Princeton: 1982.
Williams, Lt. Col. *Elephant Bill.* London: Rupert Hart-Davis, 1950.
Williamson, Captain Thomas. *Oriental Field Sports.* London: Edward Orme, 1807.
Wilson, Lt. Col. Alban. *Sport and Service in Assam and Elsewhere.* London: Hutchinson and Co. Publishers Ltd., 1924.
Wilson, Rt. Hon. Sir Guy Fleetwood. *Letters to Nobody.* New York: E. P. Dutton & Co., 1921.
Wise, Michael. *True Tales of British India.* In Print Publishing, 1993.
Woodyatt, Major-General Nigel C. B. *My Sporting Memories.* London: Herbert Jenkins Limited, 1922.
Worswick, Clark. *Princely India.* London: Thames & Hudson, 1980.
Wykes, Alan. *Snake Man: The Story of C. J. P. Ionides.* London: Hamish Hamilton, 1960.
Yadav, Ajay Singh. *The Man-Eating Wolves of Ashta.* New Delhi: Srishti Publishers and Distributors, 2000.
Zeuner, F. E. *A History of Domesticated Animals.* London: Harper & Row, 1963.
Zubrzycki, John. *The Last Nizam.* London: Macmillan & Co. Ltd., 2007..

INDEX

IRRESISTIBLE IMAGERY AND POWER: THE LION KING

As I concluded this book, it was clear to me that the images of lions had traversed the world while carrying on their back, beliefs and faiths in a diversity of cultures that connected the animal to people across the world. The lion became a symbol of royalty and kings willed it into existence around them, whether live or in art that portrayed it as a guardian of the sacred. I spent two days discussing this issue with Ruth Padel* who came up with a really succinct bit of prose which I quote:

In every culture, humans define themselves by non-humans. When they overpower an animal, they acquire, or seem to themselves to acquire, its qualities. Killing the strongest, most feared animal, you show yourself to be the strongest, most feared of men. Heracles, the greatest Greek hero, wears the skin of the lion he killed.

For most of history, divinity too has expressed itself through animals, expressing its strength and radiance through the strongest, brightest animals, like the lion. The Babylonian goddess Ishtar stands on the back of lions to bring light to earth [just] as Durga, Indian goddess of war who defeats evil and brings

---

*Ruth Padel is the author of *In and Out of the Mind: Greek Images of the Tragic Self*, *Where the Serpent Lives*, *Darwin – A Life in Poems* and *Tigers in Red Weather.*

light, rides on a lion's back. Buddhism spread in India through the roar of the lion. In Christianity, the lion symbolizes God and Christ. Assyrian seals of 2000 BCE show a human figure, probably representing a god, conquering a lion. Overpowering the lion, human becomes divine. In the Old Testament, Samson, "Man of the Sun" (sometimes read as an ancient solar divinity), possesses supernatural strength. Like Heracles, he kills a lion bare-handed. His strength is in his hair: those leonine locks are an animal token of his god-given power.

Lions evolved in Africa. So did homo sapiens. This predator with hair, like yet also unlike a man's, a mane which might also be beard, crown or sun-like halo radiating rays, is the most potent hieroglyph of human royalty and strength on the one hand, divine power and radiance on the other. Flowing through the history of nearly all continents from Africa to Assyria, Babylon, the near and far East, Eurasia, South America and India, lion symbolism percolates into all religions as [the] supreme sign of divinity. But you do not have to see wild lions to use them as a symbol. Lion imagery flourishes even where no lions naturally exist. It is everywhere in the Bible, but no lion bones have been found in Palestine and the question has been asked: were there really ever lions there? The British and Belgians symbolized their empires by a lion but in the fossil record, any sign of lions in Europe stops well before human beings evolved.

The lion answers an apparently universal human need for an animal that can symbolize simultaneously supreme human and divine power. We can never say for certain how its image migrated across the world. It may be [that] some ancestral memory of it lived on at the deepest imaginative level in all cultures. Or that when we moved from Africa into lands where no lions existed, we took the image with us to express our own most powerful self-image fused with divinity.

And wherever possible, rulers have also found ways of bringing real lions along too: as tribute, gift, beast of the chase, and a sign that their own power is also, in some sense, divine.

1. A scabbard from the Mycenaean culture, dating back to around 1500-1600 BCE. Hunting the lion was an ancient pastime.

2. A replica of the Shema seal dating back to 800-1000 BCE and said to be from Megiddo (the northern kingdoms around Israel). High ranking officials and rulers reflected their power by using lion seals.

3. A hero overcoming a lion or carrying it as a tribute in the Neo-Assyrian period (721-705 BCE).

4. Malatya (ancient Melid) in Turkey (700-800 BCE). The gods are shown overcoming a lion—a concept of depicting power that was integral to kings and rulers across the world.

5. The Assyrian and Egyptian kings began training cheetahs as early as 1700 BCE. I believe that the illustrated manuals that existed in India over the last few hundred years on the upkeep and training of cheetahs reflect the fact that there was no tradition of taming and training them in India and for those who could not read, these manuals were an essential aid to maintaining an alien animal.

6. This replica is the tribute of lions made to King Darius in the sixth century BCE and appears in Persepolis at the Great Apadana Palace. I believe it was such gifts that created captive populations of lions that were bred and carefully maintained to be released for the hunt in private hunting parks—a tradition that continued in other parts of the world including India till the twentieth century.

7. This frieze is from the Siphnian Treasury at Delphi, Greece, and depicts a lion pulling the chariot of Goddess Cybele and tearing down a giant (525 BCE).

8. The pillars of Persepolis, the ceremonial capital of the Achaemenid Empire in 550-330 BCE, appear to be a major influence on the architectural style of the Ashokan pillars of the third century BCE in India.

9. An early seal of the goddess Ishtar from more than 2,500 years ago that reflects the goddess of love and war being associated with the lion.

10. The lion was the great symbol and guardian of goddess Ishtar and is seen here at the Ishtar Gate (600 BCE).

11. From the eleventh to thirteenth centuries BCE, the goddess Ishtar may have been known in Pharaonic times as Qadesh as she stands on a lion in the midst of Egyptian gods in the Egyptian Museum at Turin. The Sumerian goddess, Nana, and the Persian goddess, Anahita, were also connected to lions.

12. Goddess Ishtar and her lion consorts or steeds resemble the depiction of Durga with her vehicle, the lion. When did the first imagery of Durga reflect itself in India? Was Ishtar a more ancient concept that travelled across regions imbibing local flavours? It is incredible that thousands of years ago such themes found ways to enter different societies' religions and beliefs.

13. Goddess Durga is shown standing on her steed, the lion, in this nineteenth-century representation, which is similar to the depiction of Goddess Ishtar 2,500 years ago. Did different cultures evoke such imagery or did such themes migrate across the world?

14 In this nineteenth-century Kalighat painting, Goddess Durga stands on her lion to defeat Mahishasura and the forces of evil. Different religions and beliefs for thousands of years connected the feminine force of the goddess to the lion.

15. Goddess Cybele with her lion chariot in the city of Madrid.

16. Cybele's cult was very active between the fourth and seventh century BCE and this theme expands with Alexander the Great's campaigns and conquests (Pergamon Altar, second century BCE).

17. Notice the similarities between the goddess Durga riding the lion more than a thousand years later in this rock relief from Mahabalipuram in southern India and the depiction of Cybele (seventh to eighth century CE).

18. A mosaic from Pella, Greece, of Alexander hunting lions which he was always shown to overcome (fourth century BCE).

19. While the women that were portrayed with the lion were goddesses, the men were in the role of deriving the power and strength of the lion by overcoming it. This relief from the Roman national museum is a depiction of the widespread concept of the hero triumphing over the lion.

20. Roman gladiators fought tigers and lions and nearly 1,000 lions were imported from Africa for these encounters between the first and third centuries CE.

21. In Biblical times, the imagery of Samson wrestling with the lion was frequently seen. Such depictions are prevalent all over the world.

22. Lions were the guardians of sacred sites across the world. Here, lions guard the roof of the Kailash temple at Ellora (eight century CE).

23. The portrayals of some of these lions (like this one at Ellora) reflect their dog-like style, as if those who sculpted them were not familiar with them.

24. The Devi or the goddess was the strong feminine force that connected with the lion, as seen in this image from the Sun Temple in Konark dating back to the thirteenth century CE.

25. Just like these lions guarded the caves at Ellora, in other places like China (to which the lion was never indigenous) they guarded sacred sites and were important tributes in history, because of which they got sculpted or painted.

26. The trade in cheetahs made an impact on textiles and this unusual piece—probably from Egypt—reflects what I think are chained cheetahs (tenth to twelfth century CE).

27. From the seventh to the seventeenth century, coursing with cheetahs was a fashion and endless gifts and tributes of cheetahs were made to rulers across the world, including in Europe and China.

28. In India, the practice of carrying cheetahs on horseback gave way to bullock carts which helped in camouflaging the cheetah from its quarry since antelope and deer were less skittish around livestock.

29. In India, most of the wealthy kept stables of cheetahs to pursue the sport of coursing with them. It was a fashion that continued till the twentieth century.

30. The sun and the lion featured in imagery that was used in several parts of the world and, until recently, was used by many kingdoms including Jamnagar in India.

31. The caption to this engraving reads: 'An Arab and negro in the service of the Guickwar [ruler of Baroda].' Africans were brought to India by sea every year, whether to serve as trainers for horses, cheetahs and lions or just as slaves and bodyguards. Hundreds of thousands, if not millions, in the last 1,000 years settled down in India and, over time, imbued our culture with their traditions.

www.ingramcontent.com/pod-product-compliance
Lightning Source LLC
Chambersburg PA
CBHW030808310726
48980CB00006B/425/J

* 9 7 8 9 3 8 2 2 7 7 5 5 2 *